Masonry Design

Masonry is found extensively in construction throughout the world. It is economical and strong. *Masonry Design*—part of the *Architect's Guidebooks to Structures* series—presents the fundamentals in an accessible fashion through beautiful illustrations, simple and complete examples, and from the perspective of practicing professionals with hundreds of projects under their belt and decades of teaching experience.

Masonry Design provides the student with and reminds the practitioner of fundamental masonry design principles. Beginning with an intriguing case study of the Mesa Verde National Park visitor center, the subsequent chapters present the fundamentals of masonry design, bending, shear, compression design, wind and seismic design, and connection design. It is a refreshing change in textbooks for architectural materials courses and is an indispensable reference for practicing architects.

Paul W. McMullin is a structural engineer and founding partner at Ingenium Design. He is also an adjunct professor at the University of Utah in Salt Lake City, Utah, USA.

Jonathan S. Price is a structural engineer specializing in historic preservation. He has a private engineering consulting practice in Pennsylvania and is registered in ten states.

Architect's Guidebooks to Structures

The Architect's Guidebooks to Structures series addresses key concepts in structures to help you understand and incorporate structural elements into your work. The series covers a wide range of principles, beginning with a detailed overview of structural systems, material selection, and processes in *Introduction to Structures*; following with topics such as *Concrete Design*, *Special Structures Topics*, *Timber Design*, and *Steel Design* to equip you with the basics to design key elements with these materials and present you with information on geotechnical considerations, retrofit, blast, cladding design, vibration, and sustainability.

Designed as quick reference materials, the Architect's Guidebooks to Structures titles will provide architecture students and professionals with the key knowledge necessary to understand and design structures. Each book includes imperial and metric units, rules of thumb, clear design examples, worked problems, discussions on the practical aspects of designs, and preliminary member selection tables, all in a handy, portable size.

Read more in the series blog: https://architectsguidestructures.wordpress.com/

Concrete Design
Paul W. McMullin, Jonathan S. Price, and Esra Hasanbas Persellin

Introduction to Structures
Paul W. McMullin and Jonathan S. Price

Timber Design
Paul W. McMullin and Jonathan S. Price

Special Structures Topics
Paul W. McMullin, Jonathan S. Price, and Sarah Simchuk

Steel Design
Paul W. McMullin, Jonathan S. Price, and Richard T. Seelos

Masonry Design
Paul W. McMullin and Jonathan S. Price

Masonry Design

Edited by Paul W. McMullin and Jonathan S. Price

NEW YORK AND LONDON

First published 2019
by Routledge
52 Vanderbilt Avenue, New York, NY 10017

and by Routledge
2 Park Square, Milton Park, Abingdon, Oxon, OX14 4RN

Routledge is an imprint of the Taylor & Francis Group, an informa business

© 2019 Taylor & Francis

The right of Paul W. McMullin and Jonathan S. Price to be identified as the authors of the editorial material, and of the authors for their individual chapters, has been asserted in accordance with sections 77 and 78 of the Copyright, Designs and Patents Act 1988.

All rights reserved. No part of this book may be reprinted or reproduced or utilised in any form or by any electronic, mechanical, or other means, now known or hereafter invented, including photocopying and recording, or in any information storage or retrieval system, without permission in writing from the publishers.

Trademark notice: Product or corporate names may be trademarks or registered trademarks, and are used only for identification and explanation without intent to infringe.

Library of Congress Cataloging-in-Publication Data
A catalog record for this title has been requested

ISBN: 978-1-138-83096-7 (hbk)
ISBN: 978-1-138-83097-4 (pbk)
ISBN: 978-1-315-73686-0 (ebk)

Typeset in Calvert MT
by Servis Filmsetting Ltd, Stockport, Cheshire

Dedication

For our professors
David Hoeppner, Torch Elliott, Larry Reaveley
Ahmad A. Hamid, Aspasia Zerva
And the gang at Wave Products
Steve, Helen, Dano, Theresa

Contents

List of Illustrations	xi
Acknowledgments	xix
List of Contributors	xxi
Introduction	xxiii
Note on the Text	xxv

1 Mesa Verde National Park Visitor Center   **1**

1.1 Project Introduction	2
1.2 Preliminary and Schematic Design Phases	6
1.3 Design Development Phase	8
1.4 Construction Document Phase	12
1.5 Construction Administration Phase	12
1.6 Building Systems	12
1.7 Project Perspectives	14

2 Masonry Fundamentals   **15**

2.1 Historical Overview	16
2.2 Codes	17
2.3 Materials	21
2.4 Material Behavior	39
2.5 Structural Configuration	47
2.6 Section Properties	48
2.7 Construction	52
2.8 Quality Control	52

	2.9 Development Length Example	53
	2.10 Where We Go from Here	54

3 Empirical Design — 56
3.1 Early Codes and Historical Precedents — 57
3.2 Evolution of the Empirical Design Method — 57
3.3 Current Code Provisions — 59
3.4 Empirical Design Examples — 63
3.5 Conclusions — 73

4 Masonry Bending — 76
4.1 Stability — 77
4.2 Capacity — 78
4.3 Demand versus Capacity — 86
4.4 Deflection — 86
4.5 Detailing Considerations — 89
4.6 Design Example — 91
4.7 Where We Go from Here — 99

5 Masonry Shear — 100
5.1 Stability — 101
5.2 Capacity — 102
5.3 Demand versus Capacity — 107
5.4 Deflection — 107
5.5 Detailing — 107
5.6 Design Example — 109
5.7 Where We Go from Here — 113

6 Masonry Compression — 114
6.1 Stability — 115
6.2 Capacity — 116

	6.3 Demand versus Capacity	121
	6.4 Deflection	121
	6.5 Detailing Considerations	122
	6.6 Slender Wall Example	125
	6.7 Where We Go from Here	134
7	**Timber Lateral Design**	**136**
	7.1 Introduction	137
	7.2 Lateral Load Paths	139
	7.3 Diaphragms	141
	7.4 Shear Walls	143
	7.5 Seismic Design Considerations	146
	7.6 Where We Go from Here	153
8	**Masonry Anchorage**	**155**
	8.1 Anchor Types	156
	8.2 Failure Modes	161
	8.3 Capacity	163
	8.4 Demand versus Capacity	182
	8.5 Detailing Consideration	182
	8.6 Anchorage Example	184
	8.7 Where We Go from Here	190
	Appendix 1	**192**
	Appendix 2	**195**
	Appendix 3	**202**
	Appendix 4	**205**
	Appendix 5	**211**
	Glossary	**214**
	Bibliography	**225**
	Index	**226**

Illustrations

FIGURES

1.1	Masonry structures in Mesa Verde National Park, courtesy of ajc architects	3
1.2	Historic pottery, courtesy Wmpearl	7
1.3	Schematic design site plan, courtesy of ajc architects	8
1.4	Design development site plan, courtesy of ajc architects	9
1.5	Design development floor plan, courtesy of ajc architects	9
1.6	Design development perspective, courtesy of ajc architects	10
1.7	Stone masonry veneer prior to installation, courtesy ajc architects	10
1.8	Stone masonry veneer after installation, courtesy ajc architects	11
2.1	Anu Ziggurat and White Temple, Uruk, modern Iraq, courtesy of Teran Mitchell	16
2.2	Chartres Cathedral, c. AD 1194–1260, Chartres, France, courtesy of Teran Mitchell	17
2.3	Sagrada Familia, Antoni Gaudi, 1882–present, Barcelona, Spain, courtesy of Teran Mitchell	18
2.4	Probabilistic relationship between demand and capacity	20
2.5	Brick masonry veneer	21
2.6	Solid brick wall showing coursing variations	21
2.7	Elements of a reinforced masonry wall	22
2.8	Concrete masonry bond beam (a) prior and (b) after grouting	22

2.9	Masonry lintel block	23
2.10	Masonry column prior to grouting	23
2.11	Masonry bond types and standard dimensions	24
2.12	Mortar ready to be placed	25
2.13	Grout placement at a wall intersection in a bond beam	27
2.14	Brick weathering regions in the continental US	30
2.15	Deteriorated masonry structure in a high sulphate and chloride environment	31
2.16	Reinforcing steel showing ribs and identifying marks	31
2.17	Reinforcing mark interpretation	32
2.18	Development and lap splice length definition for straight and hooked bars	36
2.19	LENTON® mechanical reinforcing splice coupler showing (a) partially exploded view, and (b) cutaway view	39
2.20	Stress and strain relationships in a bar	40
2.21	Conceptual stress–strain curve for masonry	41
2.22	Shear stress converted to principal stresses, showing shear loads cause tension stress	42
2.23	Clay masonry expansion relationship with time	43
2.24	Shrinkage deformation versus time	43
2.25	Creep deformation versus time	43
2.26	Conceptual stress–strain curve for steel	44
2.27	Strain distribution in member with bending load	45
2.28	Strain distributions at compression controlled, beam limit, and tension-controlled conditions	45
2.29	Masonry elements in a warehouse	47
3.1	Empirical wall heights in London circa 1894, London Building Act	58
3.2	Mapped wind speeds in the continental US, after ASCE 7–10	62
3.3	Definition of W_s and W_T for vertically spanning walls, after TMS 402	64

3.4	Wall example pier dimensions	69
4.1	Masonry wall and lintel in an industrial building	77
4.2	Masonry lintel types, courtesy Teran Mitchell	77
4.3	Lateral torsional buckling in wood beam	78
4.4	Masonry bending strain and stress distribution	80
4.5	Deformed shape of (a) simply supported and (b) multi-span beams	82
4.6	Reinforcing locations in a masonry lintel	83
4.7	Typical lintel reinforcing requirements	90
4.8	Example building configuration	92
4.9	Example lintel layout	92
4.10	Example lintel cross-section	95
5.1	Shearing action like scissors	101
5.2	Shearing action from bending	101
5.3	Paper beam with layers (a) unglued, and (b) glued	102
5.4	Shear mechanisms in masonry	104
5.5	Shear force variation at points along (a) simply supported, (b) cantilever, and (c) multi-span beams	108
5.6	Shear reinforcing detailing requirements	108
5.7	Example lintel cross-section	110
6.1	Buckled plastic straw	115
6.2	Example column interaction diagram	117
6.3	Slender wall forces and geometry	118
6.4	P-delta effect in a slender wall	119
6.5	Masonry wall construction showing vertical wall and column bars	122
6.6	Confinement ties in a flush, masonry column	123
6.7	Column lengthwise bar requirements	123
6.8	Column tie requirements	124

6.9	Free body diagram for slender wall example	125
7.1	(a) Gravity load path, (b) lateral load path turned 90 degrees	138
7.2	Comparative energy absorption for high and low deformation behavior	139
7.3	Detailed lateral load path in structure	140
7.4	(a) Ineffective, and (b) effective concrete wall to wood roof connection	141
7.5	(a) Diaphragm forces and reactions, and (b) internal forces	142
7.6	Diaphragm stress distribution (a) without, and (b) with drag struts	143
7.7	Shear wall external lateral loads and internal forces	144
7.8	Shear wall shear and moment diagrams and design forces	145
7.9	Low and high energy adsorption in different materials	146
7.10	(a) Common horizontal seismic irregularities and (b) their mitigation	149
7.11	Vertical seismic irregularities showing (a) soft story and (b) in-plane discontinuities, and their mitigation	150
7.12	Reinforcing requirements for special reinforced masonry shear walls	152
7.13	Shear wall boundary element detail	153
8.1	Common masonry anchorage configurations	156
8.2	Cast-in-place and post-installed anchors	157
8.3	Headed anchor bolts and headed stud anchors prior to installation	157
8.4	Post-installed anchors prior to installation	158
8.5	Embed plate and HSA prior to grouting	158
8.6	Adhesive anchor (a) before and (b) after installation (cutaway)	161
8.7	Tension failure modes in masonry anchorage	162
8.8	Shear failure modes in masonry anchorage	163

8.9	Tension masonry failure cones where (a) l_b controls, and (b) l_e controls	166
8.10	Shear in a masonry anchor (a) parallel, and (b) perpendicular to an edge	167
8.11	Shear masonry failure cones with the force (a) perpendicular, and (b) parallel to the top of the wall	168
8.12	Supplemental anchorage reinforcing contributing connection reliability	183
8.13	Example masonry anchor connection	184
8.14	Embed plate with DBAs to carry high tension forces, (a) detail, (b) prior to installation	191

TABLES

2.1	US masonry code summary	18
2.2	Masonry strength reduction factors	20
2.3	Unit weight of completed walls	24
2.4	Masonry mortar proportion and property specifications	26
2.5	Masonry component and mortar types, for various design compressive strength f'_m	28
2.6	Masonry modulus of rupture	29
2.7	Mild reinforcing bar sizes, areas, and tension capacities for f_y=60 k/in^2 (420 MN/m^2)	33
2.8	Masonry cover requirements	35
2.9	Maximum reinforcing size for given masonry geometry	36
2.10	Tension development length, unhooked bar f_y=60 k/in^2 (420 MN/m^2)	37
2.11	Hooked bar, tension development length f_y=60 k/in^2 (420 MN/m^2)	38
2.12	Bay spacing and floor-to-floor heights for varying building types	48
2.13	Initial horizontal member sizing guide	49
2.14	Initial column sizing guide	51

2.15	Section property equations for solid grouted masonry	51
3.1	Building height limits based on wind speed	60
3.2	Wall height to thickness ratio limits	63
3.3	Allowable compressive stresses for empirical masonry design	66
4.1	Lintel lateral bracing requirements	79
4.2	Initial lintel sizing aid	85
4.3	Maximum reinforcing area in masonry lintels	87
4.4	Deflection limits for varying lengths and criteria	89
5.1	Masonry shear strength for varying lintel depths	105
5.2	Reinforcing shear strength for varying depth to spacing	106
6.1	Initial slender wall sizes	121
7.1	Seismic lateral system R factors and maximum heights	147
7.2	Drift limits for multi-story structures	148
7.3	Maximum reinforcing area in special reinforced masonry shear walls	154
8.1	Headed stud anchor dimensions	159
8.2	Mechanical properties of anchor material	160
8.3	Post-installed anchor testing values in grouted masonry	162
8.4	Bolt and HSA steel strength	164
8.5	Masonry tension strength of a single anchor	170
8.6	Masonry tension strength of two anchors	172
8.7	Masonry tension strength of four anchors	174
8.8	Masonry shear strength of a single anchor	176
8.9	Masonry shear strength of two anchors	178
8.10	Masonry shear strength of four anchors	180
A1.1	Masonry section properties	193
A1.1m	Masonry section properties	194

BOXES

3.1	Axial compression and flexure	71
4.1	Lintel sizing	84
4.2	Design steps	91
5.1	Initial lintel shear sizing	105
5.2	Design steps	109
6.1	Initial wall sizing	120
6.2	Design steps	124
8.1	Initial anchor sizing	169
8.2	Design steps	183

Acknowledgments

Like previous editions, this book wouldn't be what it is without the diligent contributions of many people. We thank Sarah Simchuk for her wonderful figures and diligent efforts. Phil Miller for reviewing each chapter. Kevin Churilla for his work on the example figures.

We thank Wendy Fuller, our commissioning editor; Julia Pollacco, our editorial assistant; Christina O'Brien, our production editor; Jane Fieldsend, our copy editor; and Richard Sanders at Florence Production. Each of you has been wonderful to work with and encouraging and helpful along the journey. We thank everyone at Routledge who produced and marketed the book.

A special thanks to our families, and those who rely on us, for being patient when we weren't around.

We are unable to fully express our gratitude to each person involved in preparing this book. It is orders of magnitude better than it would have otherwise been thanks to their contributions.

Contributors

Jill A. Jones, AIA, LEED AP, BD+C, founded ajc architects in 1991 with a focus and passion for environmentally sensitive design that embraces the contextual elements of the site. ajc architects has provided design services to the National Park Service for over 40 projects nation-wide. Jill has led the design focus through facilitating large and diverse client groups, building consensus, and developing overall design concepts that respect and respond to the site and natural setting.

Paul W. McMullin, SE, PhD, is an educator, structural engineer, and photographer. He holds degrees in mechanical and civil engineering and is a licensed engineer in numerous states. He is a founding partner of Ingenium Design, providing innovative solutions to heavy industrial facilities. Currently an adjunct professor at the University of Utah in Salt Lake City, USA, he has taught for a decade and loves bringing project-based learning to the classroom.

Jonathan S. Price, PE, LEED AP, has taken a journey over the last 40 years in building construction or design and education. Armed with a Bachelor of Architectural Engineering degree from the University of Colorado in 1977 and a Master of Science degree in Civil Engineering from Drexel University in Philadelphia in 1992, Mr. Price has worked in various capacities in a variety of design firms. He has taught structural design at Philadelphia University since 1999 and was honored with the Distinguished Adjunct Faculty Award in 2006.

Kent R. Rigby, AIA, is a licensed architect registered in the State of Utah with 44 years of architectural experience in a wide range of project types. Kent is employed at ajc architects as a senior associate, project architect, quality control architect, and specifications writer. Kent has served in various volunteer management positions for local 501c3 non-profit arts groups and is an audio recording engineer at the UAA Counterpoint Studios. Mr. Rigby is also an accomplished visual artist and has exhibited extensively in the Intermountain area and has placed art works in public and private collections internationally.

Introduction

Reinforced masonry is the most prolific building material of the centuries and is used extensively today. We find it in homes, warehouses, institutional buildings, and manufacturing facilities. It is moderately strong, and effective in enclosing large spaces for lower cost. Masonry walls can provide the finished surface, both interior and exterior, providing simplicity and durability.

This guide is designed to give the student and budding architect a foundation for successfully understanding and incorporating masonry in their designs. It builds on *Introduction to Structures* in this series, presenting the essence of what structural engineers use most when designing masonry.

If you are looking for the latest masonry trends, or to plumb the depths of technology, you're in the wrong place. If you want a book devoid of equations and legitimate engineering principles, return this book immediately and invest your money elsewhere. However, if you want a book that holds architects and engineers as intellectual equals, opening the door of masonry design, you are very much in the right place.

Yes, this book has equations. They are the language of engineering. They provide a picture of how structure changes when a variable is modified. To disregard equations is like dancing with our feet tied together.

This book is full of in-depth design examples, written the way practicing engineers design. These can be built upon by reworking examples in class with different variables. Better yet, assign small groups of students to rework the example, each with new variables. Afterward, have them present their results and discuss the trends and differences.

For learning assessment, consider assigning a design project. Students can use a past studio project, or a building that interests them.

The project can start with determining structural loads, continue with designing key members, and end with consideration of connection and seismic design. They can submit design calculations and sketches summarizing their work and present their designs to the class. This approach requires a basic level of performance, while allowing students to dig deeper into areas of interest. Most importantly it places calculations in context: providing an opportunity to wrestle with the iterative nature of design and experience the discomfort of learning a new language.

Our great desire is to bridge the gap between structural engineering and architecture. A gap that historically didn't exist and is unnecessarily wide today. This book is authored by practicing engineers, who understand the technical nuances and the big picture of how a masonry project goes together. We hope it opens the door for you.

Note on the Text

Glossary words are shown in **bold**. Consult the Glossary for definitions. Where appropriate, imperial calculations have been aligned to the left and the metric versions aligned to the right in shaded boxes. There are also five Appendices for reference.

Mesa Verde National Park Visitor Center

Chapter 1

Jill A. Jones and Kent R. Rigby

1.1 Project Introduction
1.2 Preliminary and Schematic Design Phases
1.3 Design Development Phase
1.4 Construction Document Phase
1.5 Construction Administration Phase
1.6 Building Systems
1.7 Project Perspectives

1.1 PROJECT INTRODUCTION

In 2003, ajc architects was awarded the design contract for a new Visitor Center and Research and Museum Collections facility at Mesa Verde National Park in Colorado. This began a long and detailed process of discovery and decision making. The first task was to become familiar with the historical and cultural aspects of the Park and its reasons for existence.

Mesa Verde was the first National Park established to protect cultural resources, including archeological sites, artifacts, and other works left by indigenous peoples. The cultures of the four corners area date back 1,400 years. Originally, the people were nomadic hunters and gatherers that lived in pit houses grouped into small villages located on mesa tops and sometimes in cliff recesses. Pit houses, dating from AD 50, were primarily built in shallow pits lined with stones, covered by thatched roofs supported by vertical timbers. As these early Puebloans (village dwellers) began to develop agriculture as a major means of subsistence and complex systems of irrigation, villages became more permanent. Surplus food led to the development of stone lined storage bins. In approximately AD 750, they began to construct houses above the ground with upright walls made of poles and mud. As a natural development of their early stone work, by AD 1000, the ancestral Puebloans had developed more complex double-coursed stone **masonry** construction techniques and began to construct multi-story buildings termed pueblos. Pit houses evolved into Kivas (enclosed stone structures used for religious purposes) centered on ceremonies such as healing rites and praying for rain and successful crops.

Between AD 1000 and AD 1300, during the Classic Period, the population may have reached several thousand people, living within the stone walls of the pueblos. Then, around AD 1200, a major cultural and population shift occurred, and people began to move back into the cliff alcoves that had sheltered their ancestors centuries before. Their highly developed stone masonry construction skills were used to build complex cliff dwellings, with rooms joined together into units of 50 or more. Sandstone was skillfully shaped into rectangular blocks mortared together with mud and sand, like those seen in Figure 1.1.

Archeologists believe that the mesa and cliff dwelling sites were likely developed for defensive reasons and protection from the elements. The mesa and cliff dwellers could not be attacked from above and the lofty

Figure 1.1 Masonry structures in Mesa Verde National Park, courtesy of ajc architects

Mesa Verde National Park Visitor Center

perches afforded a full view of enemies approaching from below. The steep approaches to the dwelling sites were easy to defend by a variety of means. The south facing cliff dwelling sites took advantage of the winter sun for passive solar heating but were shaded in the summer. The mesa top sites were abandoned as the stone constructed cliff dwelling sites continued to be developed and populated.

The ancestral Puebloans abandoned the sites by AD 1300. They left no hard evidence regarding their reason for departure. Archeologists and scientists have meticulously studied the artifacts to piece together the story but have no definitive reasons for the mass exodus. Current hypotheses cite drought, crop failures, and increasing pressures from intertribal warfare.

The cliff dwellings sat vacant for over 400 years until early explorers came into the area. Historical records tell us that in 1765 New Mexico governor Tomas Velez Cachupin directed Don Juan Maria de Rivera to lead what was possibly the first expedition of modern explorers northwest of New Mexico. By 1859, several groups of explorers had entered the region. Dr. John S. Newberry, of the San Juan Exploring Expedition, was the first known European to visit the mesa area. Although he did not record any archeological sites, the expedition was the first to officially use the name Mesa Verde.

In the 1800s, the US western exploration and expansion brought more explorers into the area. In 1874, photographer William Henry Jackson ventured into Mancos Canyon and took the first photographs of a cliff dwelling, thus introducing the public to the existence of the impressive cliff dwelling sites. In 1886, a *Denver Tribune* newspaper editor published the first public call for setting the area aside as a national park, to protect the artifacts from the "vandals of modern civilization."

Some 14 years later, between 1888 and 1892, local rancher Richard Wetherill and his brothers made archeological collecting trips into Mesa Verde to obtain artifacts, resulting in the development of at least eight individual collections, several of which were later combined and sold into four larger collections. The Colorado Historical Society subsequently purchased the first collection. On December 20, 1890, Historical Society representative Benjamin Wetherill wrote a letter to the Smithsonian expressing the idea that the area be preserved as a National Park.

In 1900, the Colorado Cliff Dwellings Association was formed with the mission to preserve the cliff dwellings and joined the growing campaign to make Mesa Verde a National Park. Between 1901 and 1905, several unsuccessful bills were introduced to Congress to create "Colorado Cliff Dwellings National Park." Finally, President Theodore Roosevelt signed legislation establishing Mesa Verde National Park on June 29, 1906, with the expressed purpose of "preserving the works of man."

Between the 1930s and 1940s, the US Civilian Conservation Corps built roads, trails, park buildings, helped with archeological excavations, and created museum space. The Wetherill Mesa Archeological Project was developed between the years of 1858 and 1965, which was the largest archeological study performed in the United States. In 1972, the Wetherill Mesa site was opened to visitors.

In 1978, Mesa Verde National Park was named by the United Nations Educational, Scientific, and Cultural Organization (UNESCO) as a World Heritage Cultural Site, gaining world attention for the important archive and artifact collection.

The year 2006 saw the 100th anniversary of the Park, celebrated by yearlong events including opening and closing ceremonies, a birthday party, special tours, and a lecture series. In addition, all American Indian human remains and associated grave goods excavated within park boundaries were reburied. This ceremonial reburial was a result of 12 years of important interaction and consultation with the 24 tribes associated with the Park. This type of interaction with the indigenous peoples is an important aspect of the Park's role and mission, as well as an indicator of the importance of having tribal input on Park decision making activities and processes.

The 24 tribes legally affiliated with the Park, Mesa Verde Foundation, Mesa Verde Museum Association, National Renewable Energy Laboratory, Smithsonian Institute, University of Pennsylvania, Colorado State University, Fort Lewis College, and Crow Canyon are additional stakeholders that were included in the programming and design process for the new facility.

ajc architects, in concert with the National Park Service (NPS) and stakeholders, developed the project objectives, vision and principles, and building design concepts and requirements throughout the extensive programming exercise, resulting in an informed and highly developed design process, culminating in a building that achieved the many goals for the project.

1.2 PRELIMINARY AND SCHEMATIC DESIGN PHASES

The preliminary and schematic design for the 7,500-square foot building was informed by the programming process and interactions with the various user groups and stakeholders. Important concepts such as building siting, entry, height, scale, visibility, mass, volume, proportion, image, form, reflection, spirituality, symbolism, acoustics, landscaping, exhibit design, transparency, light, color, and materiality were all considered during the programming process. These defining concepts were maintained throughout the design process.

A site survey, provided by the NPS, provided the basis for the Site Analysis. The survey included prevailing winds, sun angles, views, locations of Pinion Pine and Utah Juniper woodlands, natural drainage, and motor vehicle noise from State Highway 160.

A primary goal for the tribes was to provide a visual "marker" on the landscape, which could also act as a cultural symbol. Based on this, and a general site location selected by the NPS, the design team identified the final building location, footprint, and configuration.

Site design considerations included erosion control during construction and air-quality protection. The design team employed low impact site development concepts to minimize the building footprint and lessen the impact on surrounding habitat. They carefully studied grading and drainage to minimize impact to watersheds from pollutants from runoff. Roof and parking surfaces were configured to reduce urban heat island effects. Exterior lighting selections minimized night sky light pollution. Natural plants and materials restored the habitat after the construction disturbance.

The building invokes a sense of wonder in the users, a key desire of the tribes. It does this by accentuating the spiritual qualities of the site, reflecting the religious practices of their ancestors, and maintaining views to significant geographical landmarks. The building is oriented towards the southern sun, and the entry portal welcomes visitors to the world of the ancestral Puebloans.

The design team provided four alternatives for building design and developed a "Choosing by Advantages" matrix for shareholder input and option selection. The matrix included cultural perspectives, maintenance, constructability, interior spaces, collection access, and natural lighting.

The selected option provided a double curved roof over the exhibit area, symbolic of native cultures and reflective of a blanket flowing in the wind. It also provided the feel of a natural cliff overhang and evoked the geographic and natural aspects of the park. In addition, the roof added visual interest to the façade, and served to frame the view to the La Platas mountains.

The building massing and arrangement is reflective of prehistoric geometric pottery designs and features symbolic iconographic designs that were informed by decorated historic storage vessels, like those seen in Figure 1.2. The intention of the design was to bestow a sense of place and containment that is both architecturally expressive and respectful of the ancestral Puebloans and their culture. Figures 1.3 through 1.6 show selected design development drawings.

The designers selected reinforced concrete masonry as the major building material because of its **load**-bearing **capacity**, **durability**, and

Figure 1.2 Historic pottery, courtesy Wmpearl

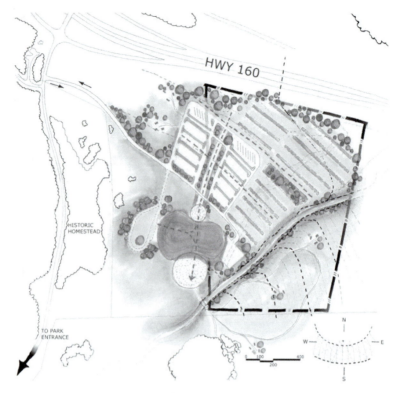

Figure 1.3 Schematic design site plan, courtesy of ajc architects

longevity. Native stone masonry **veneer** clads the exposed exterior concrete masonry units (CMU) walls, shown in Figure 1.7 and Figure 1.8, continuing the historical and indigenous context of the site. Public spaces and exposed façades employ an integrally colored, honed face CMU block, providing a natural aesthetic. In non-public spaces, a standard face CMU saves money. Careful attention to stone selection and detailing contributed to the desired reference to historic stone masonry construction at Mesa Verde.

1.3 DESIGN DEVELOPMENT PHASE

The Design Development (DD) phase refined the rooms, areas, spaces, functions, and structural, electrical, and mechanical systems.

Figure 1.4 Design development site plan, courtesy of ajc architects

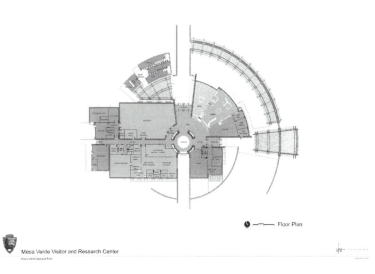

Figure 1.5 Design development floor plan, courtesy of ajc architects

Mesa Verde National Park Visitor Center

Figure 1.6 Design development perspective, courtesy of ajc architects

Figure 1.7 Stone masonry veneer prior to installation, courtesy ajc architects

Figure 1.8 Stone masonry veneer after installation, courtesy ajc architects

ajc employed extensive quality control and constructability plan checking throughout the design process. The NPS reviewed and commented on several DD drawing submittals, to which the design team responded. The NPS was very concerned with the integrity of the exterior building envelope and the interface of differing building materials and the flashing of openings and roof levels.

To provide extra protection against corrosion, the masonry ties and **anchors** were fabricated from stainless steel. Integral water repellent was added to all exterior mortar for both CMU and stone as a further measure to guard against deterioration of the exterior building materials.

ajc engaged a building envelope consultant to assist in fine tuning the building envelope. It was critical that moisture would not become trapped within the CMU and stone veneer cavity walls and create conditions that could become harmful to the museum collections. The importance of the masonry construction was never lost or diminished during Design Development.

1.4 CONSTRUCTION DOCUMENT PHASE

The **Construction Document** (CD) phase continued the building refinement and design process. The design team responded to all the NPS review comments, and they were not checked off until completed to NPS satisfaction.

ajc developed a comprehensive and dimensioned layout of the required masonry control and expansion joints to ensure masonry construction shrinkage cracking would be minimized. Exterior elevations and floor plans showed all control and expansion joints. These joints were detailed in accordance with the National Masonry Council standards.

1.5 CONSTRUCTION ADMINISTRATION PHASE

Several NPS project managers oversaw the Construction Administration (CA) phase, with assistance of the design team. All construction related information was relayed between the NPS, contractor, and design team in an established chain of command.

The original stone mason struggled with the technical requirements. The installed stone sizes and patterns did not meet the specifications, and the mortar cracked beyond acceptable standards. The NPS issued a letter of non-conforming work outlining stone masonry construction problems and the expected efforts for remediation. The original mason eventually withdrew from the project and was replaced by the contractor.

1.6 BUILDING SYSTEMS

A comprehensive geotechnical report, provided by the NPS, indicated that conventional spread footings could be utilized, bearing on a three-foot layer of moisture conditioned and compacted structural fill. This required over-excavation of the building footprint area, conditioning of the native materials, and replacement in compacted layers. To accommodate potential **slab** settlement, details separate the slab and structural **foundations**. Elastomeric sealants between slabs and foundations allow for slab movement. The slabs are 6 in thick, reinforced with a layer of welded wire mesh, and placed over a continuous vapor retarder. The exterior faces of the reinforced concrete foundation walls are treated with damp proofing to help prevent the absorption of alkalis apparent in the native soils.

The roof system consists of metal deck over steel joists and **beams**, **supported** by steel **columns** and load-bearing reinforced concrete and masonry walls. Portions of the Visitor Information Center required special arched roof joists.

Metal decking on the roof acts as horizontal **diaphragms**. Continuity across diaphragm steps is provided by **braces**. The vertical **elements** of the lateral force resisting system consist of special **reinforced masonry** and concrete **shear walls**, and ordinary steel **moment frames**. The lateral forces are delivered into the reinforced concrete foundation system by the lateral force resisting system. The foundation wall system was designed for both overturning and sliding forces. The structural design was in accordance with the 2006 International Building Code and was classified as a Category II building.

Landscape design honored and respected the Ancestral Puebloan history and natural landscape. It provided opportunities for education, research, and inspiration, while protecting valuable cultural resources. The 24 tribes associated with the site all contributed to the landscape design concepts as well as building design.

Visitors arrive at the site with spectacular views to the south of Point Lookout and east to the Las Platas mountains. Public parking is provided in two large lots north of the site. Wide welcoming sidewalks lead visitors to an arrival plaza that establishes the northern end of an important symbolic axis that extends across the site and through the building. The entry plaza provides a place for small groups of visitors to gather before proceeding further into the site.

The approach to the facility provides visitors with a zone of transition, time to decompress and prepare themselves for the experience ahead. The entry sidewalk is embraced by low walls on both sides that subtly direct visitors to the entrance court. The entry court features interpretive displays, seating areas, and quiet places for contemplation and enjoyment.

Visitors have the option of proceeding into the Visitor Information Center, the Research and Museum Collection Facility, or along an exterior interpretive walk. The walk wraps around the northeastern side of the building and connects to a courtyard that looks out to the Las Platas Mountains beyond.

From the interpretive walk and courtyard, visitors can access an informal loop trail that delicately traces through a small landscape area complete with supporting interpretive signage describing the distinctive regional landscape. The landscape immediately surrounding the facility blends naturally into the surrounding native landscape with a mix of planted native shrubs, perennials, and grasses.

The mechanical system consists of a water-to-water heat pump providing heated and chilled water to the Visitor and Information Center and restrooms via a radiant floor slab. The radiant floor slab also provides some cooling. Additional cooling is provided by active chilled beams. A rooftop fan coil provides ventilation.

The team designed the building to a LEED Gold certification level. LEED earns points for a variety of sustainability improvements, with the intent to reduce the environmental impact of the building.

1.7 PROJECT PERSPECTIVES

Prior to the opening of the new Visitor Center and Research and Museum Collection Facility, in December of 2012, the bulk of the Mesa Verde archeological collection had been housed in the "tin shed," constructed in the 1950s without insulation or climate control equipment. The new facility provides state of the art research, lab, study, and storage space for the over three million artifacts and archives in the collection. In addition, the new building serves as a visitor center for visitors from around the world, local communities and businesses, tribal members, staff, and the academic community.

The project was successfully completed due to the careful programming work and the efforts of the design team to identify the opportunities and risks associated with the project, including the participation of the list of stakeholders. The project goals were successfully met, providing a well-functioning and inspiring building that will perform as intended for generations to come. The diligence of the National Park Service Project Managers assured that careful attention to detail was maintained and the General Contractor maintained a high level of quality of construction.

Today the Mesa Verde National Park, Visitor Center, and Research and Museum Collection Facility serves as an important cultural reminder of humankind's relationship to the natural world and the built environment.

Masonry Fundamentals

Chapter 2

Paul W. McMullin

2.1 Historical Overview
2.2 Codes
2.3 Materials
2.4 Material Behavior
2.5 Structural Configuration
2.6 Section Properties
2.7 Construction
2.8 Quality Control
2.9 Development Length Example
2.10 Where We Go from Here

2.1 HISTORICAL OVERVIEW

Earth, brick, and stone are among the oldest building materials. Shortly after the glaciers receded, mankind established more permanent settlements. Buildings were simple wood or stone buildings clustered together and located in defensible sites. Early stone structures were built by stacking loose rocks found near the building site and topped with a thatch or sod roof.

Stone and wood were not abundant materials in the Fertile Crescent where the first large agrarian civilizations developed. As a result, a more prevalent and flexible material was necessary to build the growing empires. This spurred the development of hand-pressed, sun-dried bricks during the Neolithic and Bronze Ages. They were made of local clays found in abundance around the river deltas of early cities. Bricks became the building material of choice due to the availability of raw materials, ease of manufacture, and use of unskilled laborers. One man rather than scores of forced laborers could easily move a single building block to construct large dwellings and palaces, such as the ziggurats of the ancient Sumerians, shown in Figure 2.1.

Masonry changed little for thousands of years until the development of arch construction and concrete. Arch vaulting allowed brick and stone to be assembled in a way to take advantage of its high compressive **strength** while avoiding its low tensile strength. The zenith of arch and vault construction is found in the stone cathedrals of the late Gothic Period of Europe. Chartres Cathedral in France (Figure 2.2) built between AD 1194 and 1260 is an early example of structural innovations in masonry construction. Flying buttresses, stone tracery, ribbed vaulting, pointed arches, and stained-glass windows were used to great architectural effect during the Gothic periods throughout Europe.

Figure 2.1 Anu Ziggurat and White Temple, Uruk, modern Iraq, courtesy of Teran Mitchell

Figure 2.2 Chartres Cathedral, c. AD 1194–1260, Chartres, France, courtesy of Teran Mitchell

Masonry continues to be used extensively in building construction throughout the world. Availability, durability, compressive strength, and inherent fire resistance make it a popular material for residential, commercial, industrial, and institutional applications. Sagrada Familia, begun in AD 1882 in Barcelona, Spain (Figure 2.3), is the seminal work by renowned Spanish architect Antoni Gaudi. It demonstrates that traditional masonry can be used to create nontraditional forms, such as the undulating façade of Art Nouveau architecture.

2.2 CODES

Three organizations work together to publish the masonry **code** in the United States: The Masonry Society (TMS), American Concrete Institute (ACI), and the American Society of Civil Engineers (ASCE). Together they develop two primary documents: *Building Code Requirements and for Masonry Structures*[1] and *Specification for Masonry Structures*.[2] These provide guidance on member design for **gravity** and **lateral loads**, materials, and construction. Table 2.1 lists the chapters of these standards.

Figure 2.3 Sagrada Familia, Antoni Gaudi, 1882–present, Barcelona, Spain, courtesy of Teran Mitchell

Table 2.1 **US masonry code summary**

Chapter	Title	Contents
Building Code Requirements for Masonry Structures, TMS 402		
1	General Requirements	Scope, drawings, special systems, and standards
2	Notation and Definitions	Symbols used and definitions
3	Quality and Construction	Quality assurance requirements and construction considerations
4	Analysis and Design	Loads, material and section properties
5	Structural Elements	Assemblies, beams, columns, pilasters, and corbels
6	Reinforcement	Reinforcing details and anchor bolts
7	Seismic Requirements	Analysis, drift, seismic design categories
8	Allowable Stress Design	**ASD** design of anchor bolts, unreinforced and reinforced masonry
9	Strength Design	Strength (LRFD) design of anchor bolts, unreinforced and reinforced masonry
10	Prestressed Masonry	Methods, tendon stress, tension, shear, deflection, anchorages, and development

Table 2.1 *continued*

11	AAC Masonry	Strength design of Autoclaved Aerated Concrete masonry
12	Veneer	Design of anchored and adhered masonry
13	Glass Unit Masonry	Panel sizes, support, expansion joints, mortar, and reinforcing
14	Partition Walls	Prescriptive design, lateral support, and anchorage
A	Empirical Design	Requirements for experience based design
B	Masonry Infills	Design requirements for masonry inside frames
C	Limit Design	Alternative design provisions for shear walls
Specification for Masonry Structures, TMS 602		
Part 1	General	Standards, systems, submittals, quality, delivery, storage, and project conditions
Part 2	Products	Mortar, grout, masonry units, reinforcing, accessories, mixing, and fabrication
Part 3	Execution	Inspection, preparation, erection, reinforcing and grout placement, quality, cleaning, and requirement checklist

Source: TMS 402–13 and 602–13

2.2.1 Strength Reduction Factors

Safety factors reduce the chance of failure. In **strength design**, we apply them to both **demand** and capacity (load and strength). Figure 2.4 illustrates the relationship between demand and capacity. **Load factors** push the demand curve to the right, while **strength reduction factors** ϕ shift the capacity curve to the left. Proportioning members is based on providing greater capacity than demand. Where the curves overlap, failure occurs. Even if the structure is designed without errors, statistically, there is a slight (0.01%) chance of failure. The curve shape in Figure 2.4 is a function of variability. Narrow curves have less deviation from average than wider curves.

You might think it would make sense to provide substantially greater capacity than demand. This would be true if no one had to pay for the materials, or our environment didn't have to support extraction

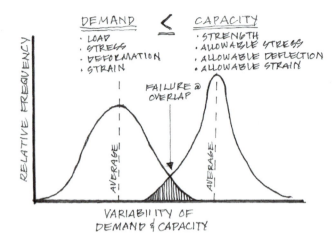

Figure 2.4 Probabilistic relationship between demand and capacity

and manufacturing costs. Since both are of concern, we balance demand with capacity to minimize risk, cost burdens, schedule, and environmental impacts. This is the art of engineering.

Strength reduction factors, or phi ϕ factors, account for variations in material quality and construction tolerances. Statistically derived from testing, they are less than one (reducing strength), with lower factors indicating greater variation. They reduce the nominal masonry design strength to ensure safe performance.

Table 2.2 summarizes strength reduction factors for masonry, which vary substantially compared to other structural materials. Anchorage reduction factors vary depending on failure mode and are further discussed in Chapter 8.

Table 2.2 Masonry strength reduction factors

Strength Mode	ϕ
Bending	0.9
Axial	0.9
Shear	0.8
Bearing	0.6
Anchorage	0.5–0.9

Source: TMS 402–13

2.3 MATERIALS

Masonry consists of bricks, mortar, **grout**, and often today, **reinforcing** steel. Working in tandem, they create durable and strong structures. We will discuss these in the following sections.

2.3.1 Masonry

2.3.1.a Masonry Units

The two major types of masonry units are brick and block. Bricks are mostly solid, and commonly used as veneer (Figure 2.5), or stacked together to create solid walls, illustrated in Figure 2.6. Block has openings that allow reinforcing steel and grout placement inside the units, shown in Figure 2.7. Varying block types allow the creation of horizontal **bond beams**, **lintels**, and columns shown in Figure 2.8, Figure 2.9, Figure 2.10, respectively.

Masonry units are made from a wide variety of materials; the two most common are clay and **cement** based. Clay is most common in bricks and cement common in block. However, there are clay blocks such as Atlas™ brick, and cement bricks. We will focus this chapter on reinforced masonry, made of concrete masonry units (CMU).

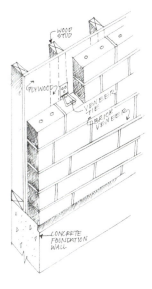

Figure 2.5 Brick masonry veneer

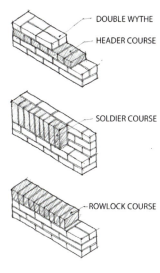

Figure 2.6 Solid brick wall showing coursing variations

Masonry Fundamentals

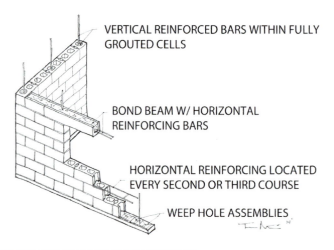

Figure 2.7 Elements of a reinforced masonry wall

Figure 2.8 Concrete masonry bond beam (a) prior and (b) after grouting

Figure 2.9 Masonry lintel block

Figure 2.10 Masonry column prior to grouting

2.3.1.b Masonry Layup

The two primary ways CMU is stacked, or laid up, are **running bond** and **stack bond**, shown in Figure 2.11. Running bond is the most common and has slightly better structural properties, because the individual blocks interlock. To accommodate the mortar joints, CMU blocks are 3/8 in less than nominal. This provides whole inch dimensions for laying out buildings in increments of 8 in.

In design, we frequently need to know the unit weight of completed masonry walls, as provided in Table 2.3.

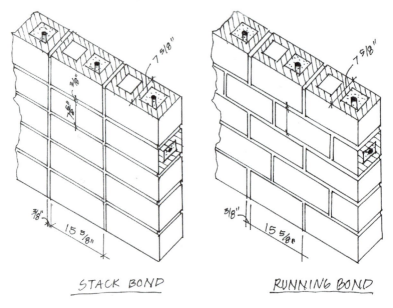

Figure 2.11 Masonry bond types and standard dimensions

Table 2.3 Unit weight of completed walls

Imperial		Wall Weights (lb/ft² of surface area)							
		Lightweight				Normalweight			
		105 lb/ft³				135 lb/ft³			
Wall Thickness (in)		6	8	10	12	6	8	10	12
Solid grouted wall		58	78	99	120	64	86	109	132
Vert Cells Grouted @	16" o.c.	41	55	68	81	47	63	78	93
	24" o.c.	35	47	58	68	41	55	68	80
	32" o.c.	32	43	53	62	39	51	62	73
	40" o.c.	31	40	49	58	37	48	59	69
	48" o.c.	30	39	47	55	36	47	57	67
No Grout		24	31	38	43	31	39	48	54

Table 2.3 *continued*

Metric		Wall Weights							
		(kN/m^2 of surface area)							
		Lightweight				Normalweight			
		$16.5\ kN/m^3$				$21\ kN/m^3$			
Wall Thickness (mm)		150	200	255	305	150	200	255	305
Solid grouted wall		2.78	3.73	4.74	5.75	3.06	4.12	5.22	6.32
Vert Cells Grouted @	406 mm o.c.	1.96	2.63	3.26	3.88	2.25	3.02	3.73	4.45
	610 mm o.c.	1.68	2.25	2.78	3.26	1.96	2.63	3.26	3.83
	813 mm o.c.	1.53	2.06	2.54	2.97	1.87	2.44	2.97	3.50
	1,016 mm o.c.	1.48	1.92	2.35	2.78	1.77	2.30	2.82	3.30
	1,219 mm o.c.	1.44	1.87	2.25	2.63	1.72	2.25	2.73	3.21
No Grout		1.15	1.48	1.82	2.06	1.48	1.87	2.30	2.59

Source: NCMA TEK Note 14–13A
1) Grout density = 140 lb/ft³ (6.7 kN/m³)
   Mortar density = 125 lb/ft³ (6.0kN/m³)

2.3.1.c Mortar

Mortar glues bricks or blocks together, shown in Figure 2.12 prior to placement. Historically, it was made of lime and sand. Today, we add cement to the mix to improve strength and durability. Sand acts as a filler and stabilizer, cement binds the sand, and lime adds workability and bond. Mortar is troweled into place between masonry units.

Figure 2.12 Mortar ready to be placed

Mortar is specified by either proportions or properties, according to ASTM C270.[3] In the US, there are four strengths of mortars: M, S, N, and O. Table 2.4 shows mortar mix proportions and property specifications. Mortar strength varies from 350 to 2,500 lb/in² (2.4 to 17.2 MN/m²).

2.3.1.d Grout

Grout fills the voids of block masonry, bonding the reinforcing steel to the block and providing additional strength. Grout is a high **slump** concrete, that is fluid enough to fill the small voids in the masonry **layup**, shown in Figure 2.13. It consists of small **aggregate**, cement, and water. Its strength ranges from 2,000 to 3,000 lb/in² (20 to 28 MN/m²), depending on the required masonry assembly strength.

Table 2.4 Masonry mortar proportion and property specifications

\multicolumn{4}{c}{Volume Method}			
Mortar	\multicolumn{3}{c}{Proportions by Volume}		
Type	Portland	\multicolumn{3}{c}{Masonry Cement}		
	Cement	M	S	N
M	1			1
M		1		
S	0.5			1
S			1	
N				1
O				1

Property (strength) Method		
Mortar	\multicolumn{2}{c}{Strength}	
Type	lb/in²	(MN/m²)
M	2,500	(17.2)
S	1,800	(12.4)
N	750	(5.2)
O	350	(2.4)

Source: TMS 602–13
1) Aggregate proportions must be between 2¼ and 3 times the total volume of cementitious material

Figure 2.13 Grout placement at a wall intersection in a bond beam

2.3.1.e Compressive Strength

Masonry strength is a function of the composite effect of the masonry unit, mortar, and grout. TMS 602 provides two methods for specifying strength: unit strength and **prism** testing. Unit strength is most common, as it specifies the unit, mortar, and grout strength. This allows the contractor to order materials with the necessary strength, and not be concerned about testing. The unit strength method is based on prism testing of specific combinations of materials.

To specify masonry compressive strength, it is best to include the design strength f'_m. If we are using the prism testing method, we are done. If we use the unit strength method, we need to specify the mortar type, and masonry unit and grout strengths. Table 2.5 provides the mortar type and masonry unit and grout strengths for varying masonry compressive strengths. The grout strength must be at least 2,000 lb/in^2 (13.79 MN/m^2). Where the masonry strength exceeds this, it must equal the design strength f'_m.

2.3.1.f Modulus of Rupture

Though masonry has little tensile strength, it does have some. In bending members, it is important to quantify this strength to ensure a member

Table 2.5 Masonry component and mortar types, for various design compressive strength f'_m

Imperial			
Net Area	CMU Minimum		Grout
Compressive Strength, f'_m	Compressive Strength, lb/in^2 Mortar Type		Strength
(lb/in^2)	M or S	N	(lb/in^2)
1,700	—	1,900	2,000
1,900	1,900	2,350	2,000
2,000	2,000	2,650	2,000
2,250	2,600	3,400	2,250
2,500	3,250	4,350	2,500
2,750	3,900	—	2,750
3,000	4,500	—	3,000
Metric			
Net Area	CMU Minimum		Grout
Compressive Strength f'_m	Compressive Strength, MN/m^2 Mortar Type		Strength
(MN/m^2)	M or S	N	(MN/m^2)
11.72	—	13.10	13.79
13.10	13.10	16.20	13.79
13.79	13.79	18.27	13.79
15.51	17.93	23.44	15.51
17.24	22.41	29.99	17.24
18.96	26.89	—	18.96
20.68	31.03	—	20.68

Source: TMS 602–13

has sufficient reinforcing. Otherwise the member may crack at strengths beyond what the reinforcing can provide. The **stress** at which masonry cracks due to bending stress is known as the modulus of rupture. It is a function of masonry layup, grout pattern, and mortar type, as provided in Table 2.6.

Table 2.6 Masonry modulus of rupture

Stress Direction	Layup Type	Masonry Type	Modulus of Rupture f_r, lb/in² (kN/m²)				
			Mortar Type			
			Portland Cement & Lime or Mortar Cement		Masonry Cement or Air Entrained Portland Cement & Lime	
			M or S	N	M or S	N
Perpendicular to Bed Joints	Running or Not Running Bond	Solid	133 (919)	100 (690)	80 (552)	51 (349)
		Ungrouted	84 (579)	64 (441)	51 (349)	31 (211)
		Fully Grouted	163 (1,124)	158 (1,089)	153 (1,055)	145 (1,000)
Paralell to Bed Joints	Running Bond	Solid	267 (1,839)	200 (1,379)	160 (1,103)	100 (689)
		Ungrouted	167 (1,149)	127 (873)	100 (689)	64 (441)
		Fully Grouted	267 (1,839)	200 (1,379)	160 (1,103)	100 (689)
	Not Running Bond	Continuous Grout	335 (2,310)	335 (2,310)	335 (2,310)	335 (2,310)
		Other	0	0	0	0

Source: TMS 402–13

2.3.1.g Masonry Durability

Durability of brick masonry in common commercial applications is driven by weathering. Depending on the region, we specify brick as severe, moderate, or negligible weathering. Figure 2.14 provides a map of weathering regions in the US.

Concrete masonry unit standards indirectly address weathering by limiting water adsorption. As the density and cement used in a block decrease, the susceptibility to weathering increases.

Cement based masonry is not advised in applications with **sulphates** or **chlorides** in the process. The sulphates degrade the cement in the block and grout, and chlorides increase corrosion of the reinforcing steel, as shown in Figure 2.15. The porous, rough nature of the cement block traps moisture and these deleterious chemicals, accelerating the deterioration.

2.3.2 Reinforcing Steel

Reinforcing steel carries the **tension** stresses in masonry. Without it, masonry structures would be limited to gravity walls with few openings, and **compression** arches. With it, a multitude of possibilities open, including lintels, **slender** walls, and high load shear walls.

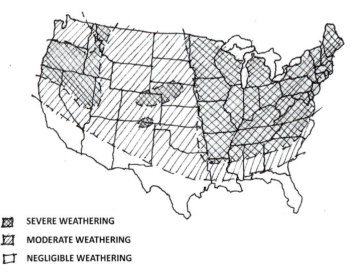

Figure 2.14 Brick weathering regions in the continental US

Figure 2.15 Deteriorated masonry structure in a high sulphate and chloride environment

Reinforcing steel is classified as mild, **prestressing**, or **post tensioning**. Mild reinforcing is synonymous with rebar. It consists of round bars with ribs to engage the grout (see Figure 2.16). It is currently available with **yield strengths** f_y of 60 and 75 k/in² (420 and 520 MN/m²). ASTM **A615** is the most common rebar. **A706** is weldable and preferred in seismic design applications where extensive yielding is anticipated as it controls the relationship between **yield** and **ultimate strengths**, ensuring ductility. TMS 402 limits the actual yield strength to $1.3f_y$.

Identifying reinforcing sizes, grades, and areas is fundamental to reinforcing construction and design. Figure 2.17 shows the reinforcing marks and how to interpret them. Table 2.7 shows available bar sizes, areas, and tension capacities for a given quantity of bars. These will help you designing with the following chapters. Note, the masonry codes limits the bar size to number 9 (M29).

Figure 2.16 Reinforcing steel showing ribs and identifying marks

Masonry Fundamentals

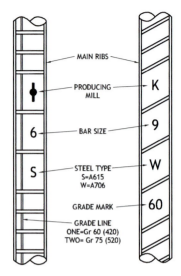

Figure 2.17 Reinforcing mark interpretation

Prestressing is another way to reinforce masonry. Strands, similar to cables, are placed in the masonry and pulled tight. This induces stress opposite the direction loads will cause. Prestressing reduces the overall volume of masonry in a wall.

2.3.2.a Reinforcing Cover

To protect reinforcing from corrosion and transfer forces between steel and masonry, the code specifies a minimum masonry covering. Clear **cover** is measured from the masonry surface to outer steel edge and presented in Table 2.8.

2.3.2.b Reinforcing Size Limits

The masonry code limits the size of reinforcing for the following reasons:

- enable more effective stress transfer by using smaller bars;
- provide shorter **development lengths**;
- limit the possibility of over-reinforcing;
- reduce grout consolidation problems.

Reinforcing steel in block masonry must meet the following requirements:

Table 2.7 Mild reinforcing bar sizes, areas, and tension capacities for $f_y = 60$ k/in² (420 MN/m²)

Imperial															Metric	
Bar Size	Dia. (in)	Bar Quantity						Bar Quantity							Size	Weight per foot (lb/ft)
		1	2	3	4	6	8	1	2	3	4	6	8			
		Area of Steel (in²)						Tension Capacity (k)								
#3	0.375	0.11	0.22	0.33	0.44	0.66	0.88	6.6	13.2	19.8	26.4	39.6	52.8	M10	0.376
#4	0.500	0.20	0.40	0.60	0.80	1.20	1.60	12.0	24.0	36.0	48.0	72.0	96.0	M13	0.668
#5	0.625	0.31	0.62	0.93	1.24	1.86	2.48	18.6	37.2	55.8	74.4	112	149	M16	1.043
#6	0.750	0.44	0.88	1.32	1.76	2.64	3.52	26.4	52.8	79.2	106	158	211	M19	1.502
#7	0.875	0.60	1.20	1.80	2.40	3.60	4.80	36.0	72.0	108	144	216	288	M22	2.044
#8	1.000	0.79	1.58	2.37	3.16	4.74	6.32	47.4	94.8	142	190	284	379	M25	2.670
#9	1.128	1.00	2.00	3.00	4.00	6.00	8.00	60.0	120	180	240	360	480	M29	3.400

Masonry Fundamentals

Table 2.7 continued

Metric

Bar Size	Dia (mm)	Area of Steel (mm²)						Tension Capacity (kN)						Imperial Size	Weight per meter (kg/m)
		1	2	3	4	6	8	1	2	3	4	6	8		
M10	9.5	71.0	142	213	284	426	568	29.8	59.6	89.4	119	179	238	#3	0.559
M13	12.7	129	258	387	516	774	1,032	54.2	108	163	217	325	434	#4	0.994
M16	15.9	200	400	600	800	1,200	1,600	84.0	168	252	336	504	672	#5	1.552
M19	19.1	284	568	852	1,135	1,703	2,271	119	238	358	477	715	954	#6	2.235
M22	22.2	387	774	1,161	1,548	2,323	3,097	163	325	488	650	975	1,301	#7	3.041
M25	25.4	510	1,019	1,529	2,039	3,058	4,077	214	428	642	856	1,284	1,713	#8	3.973
M29	28.7	645	1,290	1,935	2,581	3,871	5,161	271	542	813	1,084	1,626	2,168	#9	5.059

1) Tension capacity is unfactored, for f_y=60 k/in² (420 MN/m²)
2) For metric sizes, # is replaced by M, to avoid confusion with Imperial bar sizes

Table 2.8 Masonry cover requirements

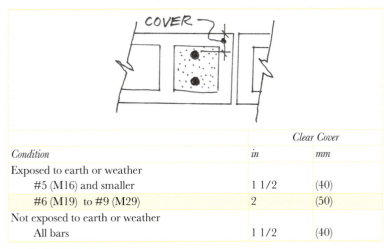

	Clear Cover	
Condition	*in*	*mm*
Exposed to earth or weather		
#5 (M16) and smaller	1 1/2	(40)
#6 (M19) to #9 (M29)	2	(50)
Not exposed to earth or weather		
All bars	1 1/2	(40)

Source: TMS 402–13

Note: 1) Cover is from masonry face to bar edge

- Bar must be smaller than
  - #9 (M29) bar
  - ⅛ the nominal member thickness
  - ¼ the least clear dimension of the cell or course
- Area must be less than 4% of the cell gross area, though 1–2% is more cost effective to build.

Table 2.9 provides maximum bar sizes for various masonry thicknesses, based on these criteria.

2.3.2.c Development Length

Reinforcing steel must be embedded into the masonry far enough, so it does not pull out, as illustrated in Figure 2.18. Known as development length, it is the distance required to develop the yield strength of the reinforcing steel. Lengths are a function of bar size, yield strength, masonry strength, bar coatings, bar **spacing**, and **cover**.

Masonry Fundamentals

Table 2.9 Maximum reinforcing size for given masonry geometry

Imperial	Nominal Masonry Thickness (in)				
	6	8	10	12	16
Max Bar Size	#6	#8	#9	#9	#9
Max A_s=4% A_{cell} (in^2)	1.80	2.44	3.08	3.72	5.00
A_s=2% A_{cell} (in^2)	0.900	1.22	1.54	1.86	2.50
A_s=1% A_{cell} (in^2)	0.450	0.610	0.770	0.930	1.25

Metric	Nominal Masonry Thickness (mm)				
	152	203	254	305	406
Max Bar Size	M19	M25	M29	M29	M29
Max A_s=4% A_{cell} (mm^2)	1,161	1,574	1,987	2,400	3,226
A_s=2% A_{cell} (mm^2)	581	787	994	1,200	1,613
A_s=1% A_{cell} (mm^2)	290	394	497	600	806

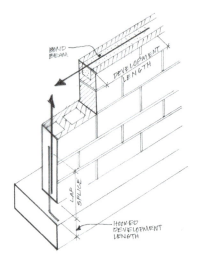

Figure 2.18 Development and lap splice length definition for straight and hooked bars

We calculate development lengths using the following:

(2.1)

$$l_d = \frac{0.13 f_y \gamma}{K\sqrt{f'_m}} d_b^2 \qquad l_d = \frac{1.57 f_y \gamma}{K\sqrt{f'_m}} d_b^2$$

where:

- f_y = bar yield strength, lb/in² (MN/m²)
- γ = 1.0 for bar size #3 (M10) to #5 (M16)
  = 1.3 for #6 (M19) to #7 (M22)
  = 1.5 for #8 (M25) to #9 (M29)
- d_b = bar diameter, in (mm)
- K = minimum of masonry cover, spacing between splices, and $9d_b$
- f'_m = masonry compressive strength, lb/in² (MN/m²)

The unhooked development length must be at least 12 in (300 mm). For epoxy coated bars, increase the length by 1.5.

For bars in tension with a standard hook, development length is:

$$l_e = 13 d_b \tag{2.2}$$

To simplify detailing, the following tables provide development lengths for:

- Tension or Compression bar without hooks—Table 2.10
- Tension with hooks—Table 2.11

Table 2.10 Tension development length, unhooked bar f_y=60 k/in² (420 MN/m²)

		Development Length, in						
Bar	d_b	Masonry Compressive Strength (lb/in²)						
Size	(in)	1,700	1,900	2,000	2,250	2,500	2,750	3,000
#3	0.375	18	17	17	16	15	14	14
#4	0.500	32	30	30	28	26	25	24
#5	0.625	45	45	45	43	41	39	38
#6	0.750	54	54	54	54	54	54	54
#7	0.875	63	63	63	63	63	63	63
#8	1.000	72	72	72	72	72	72	72
#9	1.128	82	82	82	82	82	82	82

Table 2.10 *continued*

Bar Size	d_b (mm)	Development Length, mm						
		Masonry Compressive Strength (MN/m²)						
		11.7	13.1	13.8	15.5	17.2	19.0	20.7
M10	9.53	460	440	440	410	390	360	360
M13	12.7	820	770	770	720	670	640	610
M16	15.9	1,150	1,150	1,150	1,100	1,050	1,000	970
M19	19.1	1,380	1,380	1,380	1,380	1,380	1,380	1,380
M22	22.2	1,610	1,610	1,610	1,610	1,610	1,610	1,610
M25	25.4	1,830	1,830	1,830	1,830	1,830	1,830	1,830
M29	28.7	2,090	2,090	2,090	2,090	2,090	2,090	2,090

1) For epoxy coated reinforcing, multiply values by 1.5.
2) K=1.5 for #5 (M16) bart and smaller, and 2.0 for #6 (M19) and larger.

Table 2.11 Hooked bar, tension development length f_y=60 k/in² (420 MN/m²)

| Imperial || Hooked || Metric ||
| Bar Size | d_b (in) | Development Length || Bar Size | d_b (mm) |
		(in)	(mm)		
#3	0.375	5	(130)	M10	9.53
#4	0.500	7	(170)	M13	12.7
#5	0.625	9	(210)	M16	15.9
#6	0.750	10	(250)	M19	19.1
#7	0.875	12	(290)	M22	22.2
#8	1.000	13	(340)	M25	25.4
#9	1.128	15	(380)	M29	28.7

2.3.2.d Splices

Lap splices are used to connect reinforcing where the length of continuous bars is impractical (bottom of Figure 2.18). Most splices are contact lap-type, where the bars touch, and grout binds them together. Non-contact splices are permitted when spaced less than 8 in (200 mm) or one fifth the lap length. It is prudent to stagger lap splices where reasonable, such as in walls.

Figure 2.19 LENTON® mechanical reinforcing splice coupler showing (a) partially exploded view, and (b) cutaway view. Photo is reprinted with permission and provided courtesy of ERICO International Corporation

Lap splices lengths are the same as straight bar development lengths, given in equation (2.1), and must be at least 12 in (305 mm) long.

We use mechanical or welded splices where lap lengths become too long or cause bar congestion. Mechanical splices must develop 125% of the bar yield strength and carry an International Code Council (ICC) approval report. The bolts on splice types shown in Figure 2.19 **shear** off when fully tensioned. Welded splices are permissible, but require preheating of the reinforcing, causing engineers to be wary of them.

2.4 MATERIAL BEHAVIOR

When we understand material behavior we have insight about what happens inside a structural member. Masonry behaves in a brittle (low deformation) fashion: with a linear increase in **strain**, terminating in complete loss of load carrying capability. It is over ten times stronger in compression than in tension. Steel is more ductile (withstands higher deformation before failure), linearly increasing in strain until it yields. It then follows a non-linear curve, slowly increasing in strength until it **ruptures** (separates). It has the same strength in tension and compression.

Before continuing, we must understand the terms strain and stress. Strain is the change in length divided by the original length. It is unitless, as we divide length by length. We can also think of it like percentage change in length—assuming you multiply it by 100. Stress is a measure of the internal force per unit area acting within a structural element. Stress is related to strain through the **modulus of elasticity** by the following equation, and illustrated in Figure 2.20.

$$f = E\varepsilon \tag{2.3}$$

where:

f = stress, k/in² (MN/m²)
E = modulus of elasticity, k/in² (MN/m²)
ε = strain (unitless)

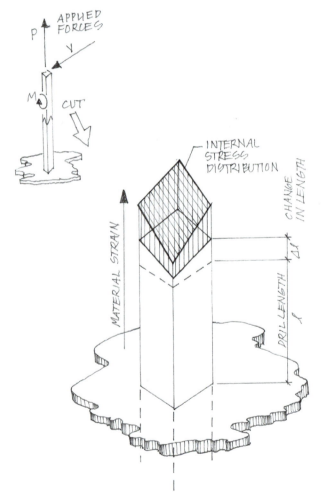

Figure 2.20 Stress and strain relationships in a bar

The elastic modulus of steel is 29,000 k/in² (200 GN/m²). For masonry, we use the following equations, differentiating between concrete and clay masonry units.

$$E_m = 900 f'_m \quad \text{for concrete masonry} \tag{2.4}$$

$$E_m = 700 f'_m \quad \text{for clay masonry}$$

where:

f'_m = is the masonry compressive strength

We use concepts of stress and strain extensively to calculate member strength and understand structural behavior.

2.4.1 Masonry

Masonry stress and strain initially follow a linear relationship under compression load, as shown in Figure 2.21. As it reaches 50% of its crushing strain the relationship changes to non-linear. After crushing, masonry quickly loses its strength.

Under shear loading, masonry cracks at 45 degrees to the applied load. This corresponds to tension, when the shear stress is converted to its principal stresses (see Figure 2.22). In other words, shear load causes tension stress—the weakest masonry behavior.

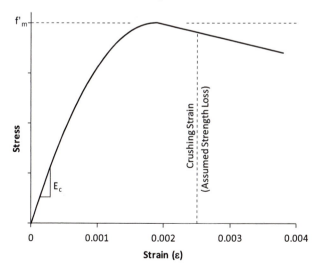

Figure 2.21 Conceptual stress–strain curve for masonry

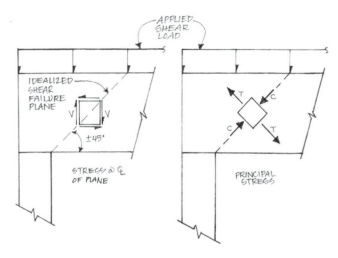

Figure 2.22 Shear stress converted to principal stresses, showing shear loads cause tension stress

2.4.1.a Volumetric Changes

Clay masonry expands as it absorbs moisture after being fired, while concrete masonry shrinks as it loses moisture as it cures.

After clay masonry is fired, it begins picking up moisture from the air or precipitation. This causes a slow, but permanent, expansion in the unit, illustrated in Figure 2.23. Expansion amounts range from 0.06% to 0.18% depending on what part of the world you are in. In the US and Canada, the expansion is closer to 0.06%. It can take upwards of five years to reach half of this expansion. While this may not seem like much, if we have a 75 ft (22.9 m) tall wall, that doesn't have relief angles, the total expansion will be:

$$l = 75 ft \left(\frac{12 in}{1 ft} \right) 0.0006$$
$$= 0.54 in$$

$$l = 22.9 m \left(\frac{1000 mm}{1 m} \right) 0.0006 \quad (2.5)$$
$$= 14 mm$$

This could be enough expansion to cause significant masonry distress.

In concrete masonry, shrinkage occurs as water leaves the block and mortar as they cure. Shrinkage rates range from 0.02 to 0.08%. It starts almost immediately after setting and continues for some time. Eventually,

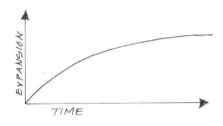

Figure 2.23 Clay masonry expansion relationship with time

it slows and then stops (Figure 2.24). For thicker members, shrinkage can last for years.

Creep is deformation due to sustained loads and occurs over an extended time period. When masonry is loaded, the absorbed water molecules flatten out, creating small changes in shape. Creep deformation increases over time, but the rate slows, seen in Figure 2.25.

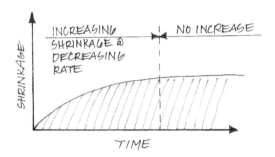

Figure 2.24 Shrinkage deformation versus time

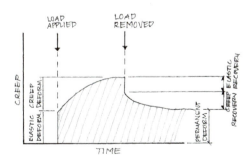

Figure 2.25 Creep deformation versus time

Masonry Fundamentals

When load is removed, the **elastic** portion of the deformation is recovered but the creep deformation remains. Reinforcing steel helps reduce shrinkage and creep deformations.

2.4.2 Steel

Stress–strain behavior of reinforcing steel consists of two primary regions: elastic and **inelastic** shown in Figure 2.26. When we remove load in the elastic range, the material returns to its original shape. If we venture into the inelastic range the material will be permanently deformed when load is removed. Actual stress-strain follows the dashed curve. We simplify the curve by assuming the steel stress is constant after yield.

2.4.3 Strength Proportions

Balancing the strength between masonry and steel is fundamental to masonry design. In beams, we need to ensure the reinforcing steel yields before the masonry crushes. This provides warning signs before a failure occurs.

The **compression controlled point** (also known as the **balanced condition**) occurs where masonry crushes at the same time the reinforcing yields. We avoid this for bending members such as beams and walls. It is unavoidable for columns, given the magnitude of compression force. Tension controlled members reach the masonry crushing point when the steel has elongated by 150%.

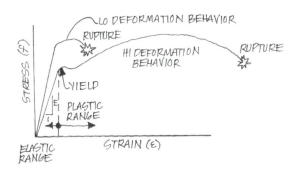

Figure 2.26 Conceptual stress–strain curve for steel

We use strain to understand how masonry and steel behave together as it varies linearly in both materials, illustrated in Figure 2.27. The relative strain in the masonry and steel determines how the member will behave: by crushing, bar yielding, or a combination of both. For the cross-section shown in Figure 2.28, we will look at three key strain

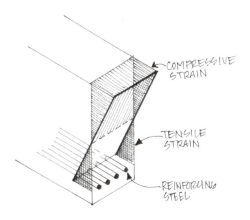

Figure 2.27 Strain distribution in member with bending load

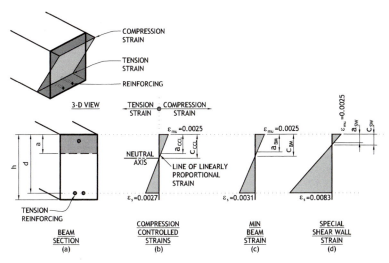

Figure 2.28 Strain distributions at compression controlled, beam limit, and tension controlled conditions

Masonry Fundamentals

relationships: balanced failure, beam, and a special reinforced shear wall. We ignore masonry tension strength in our calculations.

To help us further understand strain in masonry sections, a few definitions are in order.

- Concrete masonry crushes when it reaches a strain ε_{cu} of 0.0025 (or 0.25%), and brick masonry crushes at a strain ε_{bu} of 0.0035.
- Steel strain at the balanced condition is equal to the yield strain $\varepsilon_s = \varepsilon_y$
- Minimum steel strain ε_s in bending members is $1.5\varepsilon_y$
- Steel yield strain ε_y for 60 k/in² (420 MN/m²) reinforcing is:

$$\varepsilon_y = \frac{f_y}{E_s} \tag{2.6}$$

$$\varepsilon_y = \frac{60 k/in^2}{29,000 k/in^2} \qquad \varepsilon_y = \frac{420 kN/m^2}{200,000 kN/m^2}$$
$$= 0.0021 \qquad\qquad\qquad = 0.0021$$

We limit the bending reinforcing in a masonry section by keeping the reinforcing ratio ρ less than the maximum ρ_{max}, defined below. The reinforcing ratio is the area of reinforcing divided by the gross cross-section area. In equation form, this is:

$$\rho = \frac{A_s}{bd} \tag{2.7}$$

where:

A_s = reinforcing steel area, in² (mm²)
b = masonry width, in (mm)
d = distance from compression edge to centroid of reinforcing, in (mm)

For fully grouted members, and partially grouted walls with compression stress fully in the face shell, the limiting reinforcing ratio is given as:

$$\rho_{max} = \frac{0.64 f'_m \left(\dfrac{\varepsilon_{mu}}{\varepsilon_{mu} + \alpha \varepsilon_y} \right) - \dfrac{P}{bd}}{f_y} \tag{2.8}$$

where:

    f'_m = masonry compressive strength, k/in² (MN/m²)
    ε_{mu} = limiting masonry strain, ε_{cu} or ε_{bu} from above
    α = 1.5 for lintels and out of plane walls
       = 3.0 for intermediate shear walls
       = 4.0 for special shear walls
    P = unfactored **axial** load, k (kN)
    f_y = steel yield strength, k/in² (MN/m²)

Combining the previous two equations, and rearranging, we can find the maximum reinforcing steel area A_{max} as follows:

$$A_{max} = \frac{bd}{f_y}\left[0.64f'_m\left(\frac{\varepsilon_{mu}}{\varepsilon_{mu}+\alpha\varepsilon_y}\right) - \frac{P}{bd}\right] \qquad (2.9)$$

2.5 STRUCTURAL CONFIGURATION

Reinforced block masonry is typically limited to walls, lintels within the walls for openings, and columns. Walls are now designed as slender, where the reinforcing plays an important part of their **stability**. Figure 2.29 shows a common wall, lintel, and pilaster (column integral with the wall) in a warehouse structure.

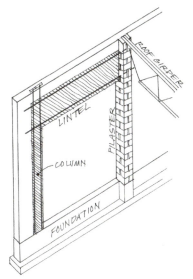

Figure 2.29 Masonry elements in a warehouse

When laying out masonry walls, stack them vertically between stories. That way the loads transfer efficiently through the masonry, and not through transfer beams. In today's age of sustainability, complexity is unjustifiable, though remarkably common.

To aid in the initial layout of a structure the following tables provide:

- typical plan dimensions and floor to floor heights—Table 2.12
- estimated depths for different horizontal members and spans—Table 2.13
- estimated size and reinforcing for columns with varying **tributary areas**—Table 2.14.

2.6 SECTION PROPERTIES

Masonry design uses area A_g, section modulus S, **moment of inertia I**, and **radius of gyration** r. They relate to strength, **stiffness**, and stability, respectively. Table 2.15 provides equations for properties of rectangular shapes. These apply to solid grouted masonry.

Table 2.12 **Bay spacing and floor-to-floor heights for varying building types**

	Plan Dimensions		*Floor to*
Building Type	*Short*	*Long*	*Floor*
	ft	*ft*	*ft*
	(m)	*(m)*	*(m)*
Hospital	25–35	30–40	15–20
	(7.5–10.5)	(9–12)	(4.5–6)
Hotel	15–30	25–35	12–15
	(4.5–9)	(7.5–10.5)	(3.5–5)
Office	25–35	30–50	13–18
	(7.5–10.5)	(9–15)	(4–5.5)
Parking	18–27	30–60	12–15
	(5.5–8.5)	(9–18)	(3.5–5)
Warehouse	20–40	35–50	18–30
	(6–12)	(10.5–15)	(5.5–9)

Table 2.13 Initial horizontal member sizing guide

System	Span/Depth	\multicolumn{10}{c}{Span (ft)}
		10	15	20	25	30	40	50	75	100	150	200	300
		\multicolumn{12}{c}{Depth (in)}
Masonry Lintel	**6.5**	18	28	37	46								
Timber Joist	**22**	6	8	12	14								
Solid Sawn Beam	**16**	8	12	16	20	24	30						
Glue-Lam Beam	**20**	6	9	12	15	18	24	30					
Truss	**12**	10	15	20	25	30	40	50	75	100			
Structural Steel Floor Beam	**20**	6	9	12	15	18	24	30	45				
Roof Joist	**24**	6	8	10	13	15	20	25	38	50			
Truss	**12**			20	25	30	40	50	75	100	150	200	300

Masonry Fundamentals

Table 2.13 continued

System		Span/Depth	Span (m)									
			3	4.5	6	7.5	9	12	15	23	30	45	60	90
			Depth (mm)											
Masonry														
Lintel		5	610	915	1,220	1,525								
Timber														
Joist		22	150	210	300	350								
Solid Sawn Beam		16	190	300	400	500	600	750						
Glue-Lam Beam		20	150	230	300	400	460	600	775					
Truss		12	150	380	500	630	760	1,010	1,300	1,900	2,500			
Structural Steel														
Floor Beam		20	150	230	300	400	450	600	750	1,150				
Roof Joist		24	150	200	250	320	380	500	640	950	1,250			
Truss		12			500	640	760	1,000	1,300	1,900	2,500	3,800	5,000	7,600

1) This table is for preliminary sizing only. Final section sizes must be calculated based on actual loading, length, and section size.

Table 2.14 Initial column sizing guide

Imperial			Metric	
Tributary Area (ft^2)	Masonry Width (in)	Number Stories for 25'–0 (7.5 m) square bay	Masonry Width (mm)	Tributary Area (m^2)
250	12	1	300	25
500	16	1	400	45
1,000	24	2	600	90
1,500	28	3	700	140
2,000	32	4	800	190

1) This table is for preliminary sizing only. Final section sizes must be calculated based on actual loading, length, and section size.
2) For normal height columns (10–12 ft, 3.0–3.7 m) and moderate loading.
3) For heavy loads, increase tributary area by 15%; for light loads, decrease area by 10%.

Table 2.15 Section property equations for solid grouted masonry

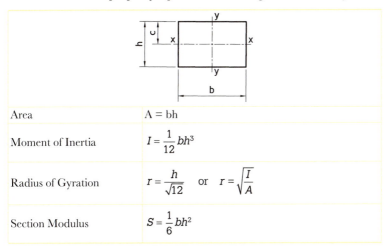

Area	$A = bh$
Moment of Inertia	$I = \dfrac{1}{12}bh^3$
Radius of Gyration	$r = \dfrac{h}{\sqrt{12}}$ or $r = \sqrt{\dfrac{I}{A}}$
Section Modulus	$S = \dfrac{1}{6}bh^2$

For partially grouted sections, we consider the openings in the block, and grout pattern. Appendix 1 provides section properties for vertically spanning masonry. We use net section properties for stress and strain, and average properties for stiffness and **deflection**. *NCMA TEK*

Note 14–1B[4] provides additional section properties for horizontally spanning masonry and non-standard unit sizes.

2.7 CONSTRUCTION

Masonry construction progresses as follows:

- Fabricating masonry units and reinforcing steel offsite.
- Placing the masonry units and associated reinforcing steel.
- Grouting the masonry.

Weather is a major consideration in masonry construction. Cold and hot temperatures, along with wind and moisture, can affect masonry strength. These all relate to the reaction between cement and water—known as hydration. Cement gains strength as it reacts with water. If it dries too soon, due to heat or wind, the chemical reaction stops, and the mortar and grout does not reach its required strength. If water freezes, its expansive **force** breaks the bond between cement and aggregate, resulting in low-strength masonry. If the masonry takes on too much moisture from the environment, it dilutes the cement, resulting in lower strength.

In hot weather, materials can be stored in the shade, and cold water used in the mix. Screens and fog sprays can reduce masonry temperature and counteract the effects of wind. In cold weather, the mortar and grout materials, along with the water, may be heated. This often is in combination with tarping and heating the space surrounding the masonry. In wet conditions, the masonry should be tarped to keep excessive moisture from the masonry. TMS 602[5] provides greater detail on these requirements.

2.8 QUALITY CONTROL

Quality control of masonry construction is critical to its successful performance and durability. Variability in quality comes primarily from block manufacturing, mortar and grout mix, and field placement. A common problem is not accounting for dimensional changes due to drying shrinkage and creep.

Masonry inspection is specified in TMS 602 and done by an impartial field technician, preferably hired by the owner, not the contractor. Quality control activities are required for:

- Mortar proportions
- Construction of mortar joints

- Grade, size, and placement of reinforcing steel
- Type, size, and placement of anchors and embedments
- Grout space, slump, and consolidation
- Cold, hot, windy, and wet weather conditions
- Size and location of structural elements
- Testing of grout and mortar.

2.8.1 Field Observations

We frequently visit the construction site to observe progress and work through challenges that arise. An engineer or architect should not direct the work. Rather, according to the contract, he or she should communicate their finding in writing to the contractor, architect, owner, and jurisdiction.

Things to look for while in the field include:

- Block. Is the masonry laid in the correct pattern? Is there clearance for the grout? Is the mortar properly bedded and tooled?
- Reinforcing Placement and Size. Are the bars in the right place and the right size? Is there excessive rust on the bars? Is the bar placed to provide the necessary cover? Is the bar tied sufficiently to keep it from moving during grout placement?
- Grout Placing. Is water being added at the site? Is the grout being vibrated into place? Is the grout being vibrated too long in one place (few seconds before being moved)?
- Protection and Curing. Is the masonry being protected and cured appropriately? Are temporary braces and shoring in place until the masonry reaches the necessary strength?

2.9 DEVELOPMENT LENGTH EXAMPLE

In this example, we calculate the plain and hooked development lengths for a #5 (M16) bar, centered in an 8 in (203 mm) wall, spaced 24 in (610 mm) apart. You can compare these with the values in Table 2.10 and Table 2.11.

Input data:

$f_y = 60,000 \dfrac{lb}{in^2}$ $f_y = 420 \dfrac{MN}{m^2}$

$\gamma = 1.0$

$d_b = \dfrac{5}{8} in$ $\qquad\qquad\qquad\qquad d_b = 15.9 mm$

$K = \min \begin{vmatrix} \text{cover} \\ \text{spacing} \\ 9d_b \end{vmatrix}$

$= \min \begin{vmatrix} (7.625 - 5/8)/2 = 3.5 in \\ 24 in \\ 9(5/8) = 5.625 in \end{vmatrix} \qquad = \min \begin{vmatrix} (203 - 15.9)/2 = 94 mm \\ 610 mm \\ 9(15.9) = 143 mm \end{vmatrix}$

The cover controls, so $K = 3.5$ in (94 mm):

$f'_m = 1{,}900 \dfrac{lb}{in^2}$ $\qquad\qquad\qquad f'_m = 13.1 \dfrac{MN}{m^2}$

Tension, plain development length:

$$l_d = \dfrac{0.13 f_y \gamma}{K\sqrt{f'_m}} d_b^2 \qquad\qquad l_d = \dfrac{1.57 f_y \gamma}{K\sqrt{f'_m}} d_b^2$$

$$= \dfrac{0.13(60{,}000\, lb/in^2)1.0}{3.5\sqrt{1{,}900}} \left(\dfrac{5}{8}\right)^2 \qquad = \dfrac{1.57(420\, MN/m^2)1.0}{94\, mm\sqrt{13.1\, MN/m^2}} (15.9\, mm)^2$$

$$= 20.0\, in \qquad\qquad\qquad\qquad\quad = 490\, mm$$

You will notice that the lengths are shorter than those in Table 2.10, as it uses a smaller K value.

Tension, hooked development length:

$l_e = 13 d_b$

$= 13 \left(\dfrac{5}{8} in\right)$ $\qquad\qquad\qquad = 13(15.9\, mm)$

$= 8.13\, in$ $\qquad\qquad\qquad\qquad\quad = 207\, mm$

2.10 WHERE WE GO FROM HERE

We now have many fundamental principles in our tool bag to expand and apply to the following chapters. We will now look at the direct design method, bending, shear, and compression member design. These are followed by a discussion on lateral design and masonry anchorage. The endnotes list additional references that provide further design guidance.[6,7]

NOTES

1. TMS, *Building Code Requirements for Masonry Structures*, TMS 402-13 (Longmont, CO: The Masonry Society, 2013).
2. TMS, *Specification for Masonry Structures*, TMS 602-13 (Longmont, CO: The Masonry Society, 2013).
3. ASTM, *Standard Specification for Mortar for Unit Masonry*, ASTM C270-14a (West Conshohocken: ASTM International, 2014).
4. NCMA, *Section Properties of Concrete Masonry Walls*, TEK 14-1B (Herndon, VA: National Concrete Masonry Association, 2007).
5. TMS, *Specification for Masonry Structures*, TMS 602-13.
6. TMS, *Masonry Designers Guide*, MDG-8 (Longmont, CO: The Masonry Society, 2013).
7. R. Drysdale, A. Hamid, *Masonry Structures: Behavior and Design*, 3rd ed. (Longmont, CO: The Masonry Society, 2008).

Empirical Design

Chapter 3

Jonathan S. Price

3.1 Early Codes and Historical Precedents
3.2 Evolution of the Empirical Design Method
3.3 Current Code Provisions
3.4 Empirical Design Examples
3.5 Conclusions

This chapter introduces the concept of **empirical design** of masonry with an emphasis on practical examples. Empirical design is a rapid method for proportioning masonry walls or as a starting point for more sophisticated methods, or as a method to produce a final design for a simple structure.

Em·pir·i·cal [əm'pirik(ə)l]: adj. based on, concerned with, or verifiable by observation or experience rather than theory or pure logic[1]

3.1 EARLY CODES AND HISTORICAL PRECEDENTS

Hammurabi's Code in c.1700 BCE Babylon comprised laws governing many types of human interaction and commerce. These included a few related to building construction, with the goal of forcing builders to be accountable for shoddy construction by requiring an eye for an eye punishment in the event of a building collapse.

An example, "If a builder build a house for some one, and does not construct it properly, and the house which he built fall in and kill its owner, then that builder shall be put to death."[2]

Hammurabi's code had no guidance for how to build, or what the loads and materials should be. Today, we have building codes that specify how much load a floor or wall must support, types of materials, allowable stresses, and fire resistance ratings.

The Empirical Design procedure specifies how thick a wall must be based on height or unsupported width without requiring analytical proof. It is, by definition, a rule of thumb.

The development and adoption of building codes is intended to protect the public against failures, injuries, and/or loss of life or injury. Whenever a momentous event occurs, such as the Great Fire of London in 1212, it becomes obvious that some rules are needed to govern construction. As a result of this fire, London banned thatched roofs. Ironically, the ban did not stop the use of thatch because in the year 1333, it was recorded that 3000 reed bundles were delivered to the Tower of London for a new roof on a new building to house the "*engines of war, mangonals, etc.*"[3]

3.2 EVOLUTION OF THE EMPIRICAL DESIGN METHOD

London had another great fire in 1666, destroying about 80% of the city, which led to the adoption of the London Building Act requiring that new walls were to be made of brick or stone.[4] In 1774, the newly adopted

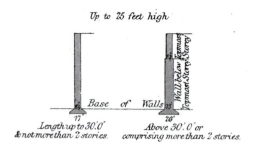
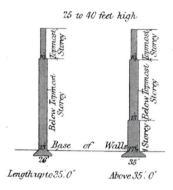

Figure 3.1 Empirical wall heights in London circa 1894.
Source: London Building Act

building code specified thickness depending on the building height. In 1894, the code specified thickness based on height and use, shown in Figure 3.1.[5]

Along a similar line, Boston developed a code to protect against fire, the first in 1631. Then in 1927, the UBC specified the minimum thickness of masonry walls beginning at 12 inches (30 cm) thick for two story structures, increasing to 24 inches (60 cm) at the base for eight story structures. Maximum h/t ratio (calculated based on the floor to floor height) was set at 10.[6]

The adoption of the American Standard Building Code Requirements for Masonry, ASA A41.1 published in 1944, specified vertical or horizontal span over thickness ratios based on mortar type, etc. Compression stress was limited then as it is now. The maximum allowed compression stress on the gross cross-section was

80 lb/in^2 (0.552 MN/m^2) assuming **Portland** based mortars. This compares favorably with the current code limit of 90 lb/in^2 (0.62 MN/m^2) assuming 1,500 lb/in^2 (10.34 MN/m^2) units and Type N mortar.

The rules of thumb contained in ASA A41.1 have been updated and now are found in the 2013 Masonry Code,[7] Appendix A, "Empirical Design of Masonry," which we will simply call empirical design. No other building element in modern construction can be proportioned simply based on empirical design limits.

3.3 CURRENT CODE PROVISIONS

3.3.1 Applicability

Empirical design is limited to buildings of limited height and in areas away from strong earthquakes or hurricanes. If you elect to use this method either for final or preliminary design, remember that some simple and straightforward calculations are required. These include a check on gross compression stress, modifications to the h/t ratio for the effects of openings, and restrictions on gravity loading.

The code requires minimum wall thickness, bonding of multi-**wythe** masonry walls, and sets limits on floor diaphragm dimensions. As an example of the latter, the length over width for a steel deck or plywood diaphragm shall not exceed 2:1. And for other stiffer materials, such as cast in place concrete or concrete on steel deck, the ratio is a bit greater.

Also, loads from beams and joists are only to be applied to the middle third of the wall or pier's thickness; this is to avoid tension within the cross-section. The middle third of the wall's thickness is the kern, a term used for any square or rectilinear compression element, masonry, walls, piers, foundations, etc. For example, because soil has no tensile strength, the engineer should proportion footings, so they are in compression across their full area.

3.3.2 Limits

The masonry Society (TMS 402 specifies maximum building height based on wind speed, provided in Table 3.1. The wind velocity limits are found in the current ASCE factored wind speed maps, shown in Figure 3.2. These mapped wind speeds exceed the maximum wind velocity by a factor-of-safety. Wind **pressure** is a function of the wind velocity squared and a 115 mph (51 m/s) factored wind speed is

Table 3.1 Building height limits based on wind speed

Element Description	Building Height, ft (m)	Basic Wind Speed, mph (mps)[a]			
		Less than or equal to 115 (51)	Over 115 (51) and less than or equal to 120 (54)	Over 120 (54) and less than or equal to 125 (56)	Over 125 (56)
Masonry elements that are part of the lateral-force-resisting system	35 (11) and less	Permitted	Permitted	Not Permitted	Not Permitted
	Over 180 (55)	Not Permitted	Not Permitted	Not Permitted	Not Permitted
Interior masonry load-bearing elements that are not part of the lateral-force-resisting system in buildings other than enclosed as defined by ASCE 7	Over 60 (18) and less than or equal to 180 (55)	Permitted	Not Permitted	Not Permitted	Not Permitted
	Over 35 (11) and less than or equal to 60 (18)	Permitted	Permitted	Not Permitted	Not Permitted
	35 (11) and less	Permitted	Permitted	Permitted	Not Permitted

Exterior masonry elements that are not part of the lateral-force-resisting system	Over 180 (55)	Not Permitted	
	Over 60 (18) and less than or equal to 180 (55)	Permitted	Not Permitted
	Over 35 (11) and less than or equal to 60 (18)	Permitted	Not Permitted
Exterior masonry elements	35 (11) and less	Permitted	Not Permitted

Note: [a]Basic wind speed as given in ASCE 7

Source: TMS 402–13

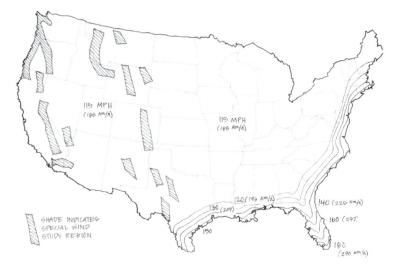

Figure 3.2 Mapped wind speeds in the continental US, after ASCE 7–10

equivalent to 90 mph (40 m/s) unfactored. When using the Allowable Stress Design load combinations, you'll see a 0.6 factor applied to the wind pressure. Prior editions of ASCE 7 mapped the unfactored wind speed and previously most of the US was in the 90 mph (40 m/s) category.

The mapped wind speed for the US interior is typically 115 mph (51 m/s). In regions where the wind speed could exceed 125 mph (56 m/s), the empirical design method is not permitted. In high seismic areas, there are similar prohibitions.

Note that older structures that are to undergo a significant renovation will normally require strengthening to meet current codes even if they had met code requirements when constructed.

3.3.3 Height to Thickness Ratios

Walls proportioned using empirical design often will be thicker than those based on an engineered approach. The typical wall height or length over thickness ratio limits for empirically designed masonry are listed in code, provided in Table 3.2. In contrast, engineered walls are

Table 3.2 **Wall height to thickness ratio limits**

Construction	Maximum l/t or h/t
Load-bearing walls	
Solid units or fully grouted	20
Other than solid units or fully grouted	18
Non-load-bearing walls exterior	18

Source: TMS 402–13

typically reinforced vertically and horizontally with either wire joint reinforcing or reinforcing bars.

3.4 EMPIRICAL DESIGN EXAMPLES

3.4.1 Tensile Stress

Do masonry walls experience tensile stress when subjected to the design wind speed limit of 115 mph (51 m/s)? There is no empirical design requirement for checking whether tension stress occurs, we are just curious. 90 mph (40 m/s) is the unfactored wind speed that is equivalent to a factored wind speed of 115 mph (51 m/s). We can convert this wind speed to a pressure using the ASCE code equation $p = 0.00256 \times V^2$ (metric equiv $0.00062 \times V^2$). This velocity results in a basic pressure of about 20 psf (0.99 kPa).

Assume you are designing a 10' (3 m) high, 8" (20 cm) thick hollow block wall. The **wind load** bending moment would be $20 \times 10^2/8 = 250$ ft-lbs (339 N-m).

Hollow 8" (20 cm) block wall has a section modulus of about 80 in^3/ft (0.0043 m^3/m) and the net area is 30 in^2/ft (0.063 m^2/m) if we assume face shell bedding only. Also, the standard type of hollow 8" (20 cm) nominal CMU weighs about 55 psf (2.63 kPa); load at mid-height is 5×55 psf = 275 plf (4.0 N/mm). Therefore, stress due to selfweight = 275/30 = about 10 psi (69 kPa). The bending stress is about 250 ft-lbs $\times$ 12 in/ft /80 in^3 = 38 psi (262 kPa) so there is a net tension stress of $38 - 10 = 28$ psi (193 kPa). Conclusion, empirically designed walls loaded with the maximum wind pressure may experience tension stress.

Sidenote: OSHA rules (see below) require bracing for walls 8' (2.4 m) or greater in height (above the footing). Some contractors either ignore or are not aware of this limit. The author's practice is to require some nominal vertical reinforcing and dowels in the foundation extending up into the wall at least 4' (1.2 m) in height to provide some degree of stability during construction when the wall is acting as a cantilever. As the engineer of record, we do not claim this reinforcing supplants the OSHA (US Occupational Safety and Health Administration) requirements because construction safety is the contractor's responsibility.[8]

"All masonry walls over 8 feet in height shall be adequately braced to prevent overturning and to prevent collapse unless the wall is adequately supported so that it will not overturn or collapse. The bracing shall remain in place until permanent supporting elements of the structure are in place."

Let's review the effects of openings on the minimum wall thickness:

Because windows and doors do not carry load, the gravity forces have to bypass the openings, and this funnels compression force to the piers between openings. Also, wind loads are carried by the piers. The empirical design provisions are modified to compensate. The procedure is to divide allowed h/t or l/t by the square root of W_T/W_S. Note the different definitions of W_s in Figure 3.3.

Example dilemma: The reader may have experienced a situation similar to the following …

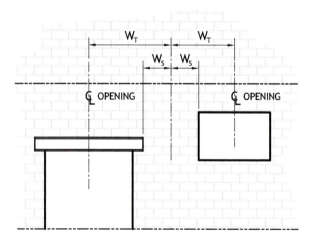

Figure 3.3 Definition of W_s and W_T for vertically spanning walls, after TMS 402

- Architect to engineer: "I'd like to cut several 6' (1.83 m) wide openings in the front elevation of a building having unreinforced CMU bearing walls but I will leave 16" (40 cm) between them. The wall is 12" (30 cm) thick wall, only 10' (3.04 m) high and just supports a roof, should be ok, right?"
- Engineer: perhaps, let me run a few quick numbers. Using an Empirical approach, the minimum wall thickness is 10' × 12/18 = 6.66" (3.04/18 = 0.17 m = 17 cm) but what is the maximum H/t ratio considering the openings?

W_T = (6'+1.33')/2 = 3.665' (SI units (1.83 m+0.4 m)/2 = 1.16 m))

W_S = 1.33'/2 = 0.66' (SI units 0.4 m/2=0.2 m)

Sqrt (3.665/0.66) = 2.36 Max permitted h/t = 18/2.36 = 7.63, therefore minimum wall thickness using empirical approach is 10' × 12/7.63 = 15.7" (40 cm) assuming we use masonry lintels, but this will be difficult to install in an existing wall, the architect isn't going to like this!

The engineer calls the architect with bad news: "your wall would need to be at least 16" (40 cm) thick to do what you want, otherwise we need to reinforce the piers or spread the openings out." Silence on the other end implies the architect may have painted a prettier picture for the owner. He considers calling another engineer he thinks is less conservative but asks: "OK, how far apart should do the openings need to be?"

- Engineer: "If we use steel lintels with 8" (20 cm) of bearing at each end the pier width has to be wider than the case where we build U-shaped lintels into the wall which means we'd have to remove the wall up to the roof. Or we can engineer the wall and add reinforcing bars."
- Architect: "OK, let's reinforce the pier, I already promised the owner we could do this with 16" piers. Also, let's assume masonry (U-shaped) grout filled reinforced lintels. The owner doesn't want to see steel."
- Engineer: "OK, to be safe, let's require the contractor temporarily support the roof and rebuild the wall above and between the openings using reinforced (engineered) CMU. Then we can be less conservative with the pier width."

Calculations are required: As noted in the example above, empirically designed masonry requires some calculations. In addition to the h/t ratio check, we must check the compression stress on the gross cross-section.

Table 3.3 Allowable compressive stresses for empirical masonry design

Construction: compressive strength of masonry unit, gross area, lb/in^2 (MN/m^2)	Allowable compressive stresses based on gross cross-sectional area, lb/in^2	
	Type M or S mortar	Type N mortar
Solid masonry of brick and other solid units of clay or shale; sand-lime or concrete brick		
8,000 (55. 16) or greater	350 (2.41)	300 (2.07)
4,500 (31.03)	225 (1.55)	200 (1.38)
2,500 (17.23)	160 (1.10)	140 (0.97)
1,500 (10.34)	115 (0.79)	100 (0.69)
Grouted masonry of clay or shale; sand-lime or concrete		
4,500 (3 1.03) or greater	225 (1.55)	200 (1.38)
2,500 (17.23)	160 (1.10)	140 (0.97)
1,500 (10.34)	115 (0.79)	100 (0.69)
Solid masonry of solid concrete masonry units		
3, 000 (20.69) or greater	225 (1.55)	200 (1.38)
2,000 (13.79)	160 (1.10)	140 (0.97)
1,200 (8.27)	115 (0.79)	100 (0.69)
Masonry of hollow load-bearing units of clay or shale		
2,000 (13.79) or greater	140 (0.97)	120 (0.83)
1,500 (10.34)	115 (0.79)	100 (0.69)
1,000 (6.90)	75 (0.52)	70 (0.48)
700 (4.83)	60 (0.41)	55 (0.38)
Masonry of hollow load-bearing concrete masonry units, up to and including 8 in (203 mm) nominal thickness		
2,000 (13.79) or greater	140 (0.97)	120 (0.83)
1,500 (10.34)	115 (0.79)	100 (0.69)
1,000 (6.90)	75 (0.52)	70 (0.48)
700 (4.83)	60 (0.41)	55 (0.38)

Table 3.3 *continued*

Masonry of hollow load-bearing concrete masonry units, greater than 8 and up to 12 in (203 to 305 mm) nominal thickness		
2,000 (13.79) or greater	125 (0.86)	110 (0.76)
1,500 (10.34)	105 (0.72)	90 (0.62)
1,000 (6.90)	65 (0.45)	60 (0.41)
700 (4.83)	55 (0.38)	50 (0.35)
Masonry of hollow load-bearing concrete masonry units, 12 in (305 mm) nominal thickness and greater		
2,000 (13.79) or greater	115 (0.79)	100 (0.69)
1,500 (10.34)	95 (0.66)	85 (0.59)
1,000 (6.90)	60 (0.41)	55 (0.38)
700 (4.83)	50 (0.35)	45 (0.31)
Multi-wythe non-composite walls		
Solid units		
2500 (17.23) or greater	160 (1.10)	140 (0.97)
1500 (10.34)	115 (0.79)	100 (0.69)
Hollow units of clay or shale	75 (0.52)	70 (0.48)
Hollow units of concrete masonry of nominal thickness,		
up to and including 8 in (203 mm)	75 (0.52)	70 (0.48)
greater than 8 and up to 12 in (203–305 mm)	70 (0.48)	65 (0.45)
12 in (305 mm) and greater	60 (0.41)	55 (0.38)
Stone ashlar masonry		
Granite	720 (4.96)	640 (4.41)
Limestone or marble	450 (3.10)	400 (2.76)
Sandstone or cast stone	360 (2.48)	320 (2.21)
Rubble stone masonry		
Coursed, rough, or random	120 (0.83)	100 (0.69)

Source: TMS 402–13

The allowed compression stress ranges from 50 psi (0.35 MPa) to 350 psi (2.41 MPa) depending on unit strength, unit type (hollow or solid), and mortar type, listed in Table 3.3. Putting these numbers in perspective, consider an 8" (20 cm) thick wall with an axial load of 4000 pounds/ft (58.4 kN/m). The stress on the gross cross-section will be $4000/(7.625 \times 12) = 42$ psi (0.29 MPa), which is less than the permitted stress for hollow block using type N mortar (type N has more lime in it and is weaker and commonly used for veneer). The same wall, if fully grouted and using type S or M mortar, would be about 3 to 4 times stronger.

Earlier editions of the masonry code, ASA A41.1 published in 1944, permitted only 70 psi on the gross cross-section for units laid with type B mortar, now called type N or 1:1:6.

3.4.2 Three-Story Example

We will plan out the floor and wall structure for a new three-story building to be constructed in Ames, Iowa. The walls are envisioned to be brick veneer over CMU, the floors and roof will be framed using engineered lumber (wooden I-joists supporting the floors and prefabricated roof trusses) with plywood sheathing and gypsum sound deadening.

Insulation will be in the cavity between the CMU and brick therefore no composite action can be assumed between the CMU and brick.

Step 1. Classify the use: residential, apartments

Step 2. Determine the applicable loads
- Floor **live load**: 40 psf (1.92 kN/m^2) for private rooms and private corridors, 100 psf (4.79 kN/m^2) for public spaces, corridors and stairs. 80 psf (3.83 kN/m^2) for corridors above the first floor.
- Roof Snow load: Ground snow load = 30 psf (1.44 kN/m^2) (ASCE 7-10 map)
- Wind Velocity: 115 mph (51 m/s) (factored) – not hardened against tornadoes

Step 3. Determine the appropriate dimensions for design:
- 24' × 60' × 10' floor to floor height and 3rd floor to roof truss bearing (7.3 m × 18.3 m × 3.05 m)

Step 4. Assume a structural system:
- Engineered wooden joists at 16" (40 cm) on center spanning the 24' (7.3 m) dimension.
- Walls: 8" (20 cm) CMU with brick veneer

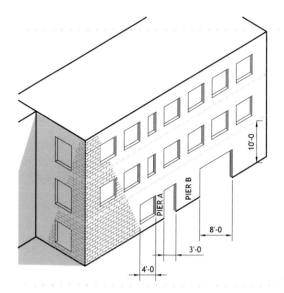

Figure 3.4 Wall example pier dimensions

Step 5. Preliminary Wall Design:
- First step is to calculate the h/t ratio (height over thickness) for the proposed CMU bearing wall – shown in Figure 3.4. Wall thickness will be verified later on by calculation. Assume 8" (20 cm) thick CMU walls will work. We typically assume a size or thickness when the element being designed is in compression. h/t = (10' × 12"/ft)/18 = 6.66" (17 cm), OK so far; 8" (20 cm) nominal exceeds this minimum.

Step 6. Check Diaphragm length: Maximum length / width of building = 60'/24' = 2.5 > 2.0 for a flexible diaphragm, no good. We need an interior shear wall and we'll get to that later.

Step 7. DESIGN PIER A

Check the effect of openings: Proposed 4' (1.22 m) wide window adjacent to 3' (0.91 m) wide door. What width pier is required between the two openings? Assume 8" (0.2 m) hollow CMU with U-shaped masonry lintels over the openings.

Empirical Design

Actual h/t = 10' × 12"/ft/8 = 15 (3.05 m/0.2 m = 15). Set this against the adjusted maximum h/t accounting for openings to obtain 15 = 18/sqrt(W_T/W_S), thus (sqrt(W_T/W_S))^2 = (18/15)^2.

Therefore, W_T/W_S = 1.44, also W_T = 4/2+W_S, (1.22 m/2 + W_S).

Substituting 1.44 W_S for W_T in the second expression above, W_S = 2/0.44 = 4.54' (0.61/0.44=1.39 m), thus the pier width must be 2 × 4.54' = 9.08' (2.77 m), let's say 9' (2.75 m). Refer to Figure 3.4.

Checking, W_T/W_S = (4/2 + 4.54)/4.54 = 1.44; and max h/t = 18/sqrt(1.44) = 15.

SI units: (W_T/W_S = (1.22/2 + 1.38)/1.38 = 1.44; and max h/t = 18/sqrt(1.44) = 15).

STEP 8. DESIGN PIER B

Using the same approach as for pier B we find the minimum pier width is about 18' (5.5 m).

We know that the architect won't be happy with a 9' (2.74 m) pier at A and twice that at pier B so we must either engineer and reinforce the piers, or increase the wall thickness, or a combination of both.

Step 9. Reproportion the wall: Let's thicken the wall to 12" (30 cm) and recalculate the pier width. It pays to develop a spreadsheet to make the work go faster. By thickening the wall to 12" (0.3 m), we determine Pier A can be reduced to 1.8' (0.55 m), but let's say 2' (0.6 m) and pier B can be 3.6' (1.1 m) but let's use 4'-0" (1.22 m).

Step 10. Wall Compression check: Assume the joists span the full 24' (7.3 m) width, then the load on the walls would be 24/2 × (2 floors) (25 psf DL + 40 PSF Live) + 24/2 × (15 psf DL + 30 psf Snow) + wall weight.

(7.3 m/2 × 2 floors) (1.2 kN/m^2 DL + 1.9 kN/m^2 Live) + 7.3 m/2 × (0.72 kN/m^2 + 1.44 kN/m^2 Snow) + wall weight))

The wall height is 30' (9.14 m), and 10" (25 cm) hollow CMU weighs about 60 psf (2.87 kN/m^2) but let's assume 70 psf (3.35 kN/m^2) anticipating that we may grout some cells. Total load on walls at base of first story would be 12 × 2 × 65+ 12 × 45+ 30 × 70 = 4200 plf

SI units: 3.66 m × 2 × 3.11 kN/m^2 + 3.66 m × 1.44 kN/m^2 + 9.14 m × 3.35 kN/m^2 = 58.65 kN/m

Compression on gross area = 4200plf/(9.625" × 12") = 36.4 psi < 75 psi (58.65 kN/m/0.244 m = 0.24 MPa < 0.52 MPa) allowable using

Type S mortar, therefore OK. This check is inconclusive because there are wall openings that magnify the stress on piers. Assuming pier B is 4' (1.22 m) wide, it must resist 4200 × (8/2+4+3/2)=39,900 pounds (58.65 kN/m × (2.44 m/2+1.22 m+0.91 m/2)=170 kN) over a gross area of 10" × 48" (0.25 m × 1.22 m) then the compression stress would be about 39,900#/480in^2 = 83 psi (170 kN/(0.25 × 1.22 = 0.56 MPa) which exceeds the permitted value (see Table 3.3). So, we are in trouble on two fronts, even with a 10" (25 cm) wall, the stress is too great and the min pier width will not be easily satisfied.

The applicable equations that are to be used in the case of an engineering analysis can be found in chapter 8 of the *Building Code Requirements and Specification for Masonry Structures*[9] (note: equations are numbered as per their source):

AXIAL COMPRESSION AND FLEXURE

8.2.4.1 Axial and flexural compression

Members subjected to axial compression, **flexure**, or to combined axial compression and flexure shall be designed to satisfy Equation 8-14 and Equation 8-15.

$$\frac{f_a}{F_a} + \frac{f_b}{F_b} \leq 1 \tag{8-14}$$

$$P \leq \left(\frac{1}{4}\right) P_e \tag{8-15}$$

where:

(a) For members having an h/r ratio not greater than 99:

$$F_a = \left(\frac{1}{4}\right) f'_m \left[1 - \left(\frac{h}{140r}\right)^2\right] \tag{8-16}$$

(b) For members having an h/r ratio greater than 99:

$$F_a = \left(\frac{1}{4}\right) f'_m \left(\frac{70r}{h}\right)^2 \tag{8-17}$$

(c) $F_b = \left(\frac{1}{3}\right) f'_m$ \hfill (8-18)

(d) $P_e = \dfrac{\pi^2 E_m I_n}{h^2} \left(1 - 0.577 \dfrac{e}{r}\right)^3$ \hfill (8-19)

Empirical Design

Conclusion: We need to use an engineered design and we will combine the effects of axial compression and flexural bending stresses using the unity equation 8-14. See the following code guidance but first some terminology: Fa is actual stress, Fb is the allowable, fa actual compression stress due to axial load and Fa the allowable stress, h is the height, r is the radius of gyration, and f'_m is the specified compression strength.

A quick check of the axial load on a hollow wall yields the following results. Let's assume we are trying a 10" (25 cm) CMU wall. The face shell area = $1.375 \times 2 \times 12 = 33$ sq in per foot (SI 35 mm × 2 × 1000 mm=70,000 mm²/meter). Let's also assume f'_m=1500 psi (10.34 MPa), a reasonable value for modern masonry.

I= $(9.625^3-(9.625-2.75)^3) \times 12"/12 = 567$ in⁴, therefore r=sqrt(567/33)=4.14,

SI units I=$(244^3-(244-70)^3) \times 1000/12$=771E6 mm⁴, therefore r = sqrt(771E6/70,000)=105 mm

S=567/(9.625/2)=117 in³

SI units S=771E6/(244/2)=6.3E6 mm³=0.0063 m³

and h/r=120/4.14=29<99 therefore use ACI 530–13 eqn. 8-16

Fa=¼ × 1500 × [1−(120/(140 × 4.14))2]=359 psi (2.48 MPa)

Actual stress on the wall is (39,900/4')/33=300 psi<359 psi,

SI units 170 kN/(1.22 × 70,000 mm²)=1.99 MPa<2.48 MPa OK so far.

Next step, calculate wind load and bending on the piers, calculate the reinforcing required if the walls go into tension due to bending. If there's no tension due to bending (if the compression exceeds bending stress), we may use equation 8-14 from the ACI 530-13 code:

Bending stress assuming 20 psf (0.96 kN/m²) wind and the openings shown above with pier A at 2' (0.61 m) and pier B at 4' (1.22 m). The critical case would be pier B but let's check both. First check pier B assuming 20 psf (0.96 kN/m²) lateral wind pressure. Accounting for the openings, the wind load on pier B would be (8/2+4+3/2) × 20=190 plf (2.77 kN/m). Assuming pinned:pinned supports and a vertical span of 10', the moment M=190plf × 10'²/8=2375 ft lbs (35 kN/m). Spreading that out over 4' (1.22 m), we find the moment is about 600 ft/lbs per foot (2.7 kN/m/m). Bending stress therefore is 600 × 12/118=61 psi (2.7/0.0063=0.428 MPa)

Allowable bending stress is 0.33 × f′m=500 psi (3.45 MPa) assuming f′m=1500 psi (10.34 MPa).

Now, using equation 8-14, does fa/Fa+fb/Fb<1.0?

300/359+61/500=0.96<1.0. (1.99/2.48+0.428/3.45=0.93<1.0) it works!

Let's check the 2' (0.61 m) wide pier A, actual axial stress = 4200 plf × (3/2+2+4/2)/2/33=350 psi (2.4 MPa). The allowable is 359 psi (2.475 MPa) so there's not much left over to resist wind loads. Let's grout that pier and any pier less than or equal to 4' (1.22 m).

A final check on pier A is needed assuming fully grouted condition.

Plan on using a 10" (25 cm) thick CMU bearing walls with grouted and reinforced piers between openings. Finalize the design using the engineered approach noted above.

Check that floor loads are delivered to the wall in a way that does not induce a lot of bending by pocketing beams into the wall and anchoring them to an embedded plate.

We don't want to go down a blind alley either by checking and forcing an 8" (20 cm) wall to work where it really ought to be a 10" (25 cm) or even a 12" (30 cm) thick wall. This dilemma is one of the biggest in our profession. Spending more time to make the project better saves the owner money and construction cost at the expense of design fee. A loose set of drawings leads to field interpretations made by those who may not know your design intent and potential claims due to errors and omissions. The best approach in the early stages of design is to spend some time to examine various options and make informed design decisions.

3.5 CONCLUSIONS

In planning a new masonry bearing wall project whether it is engineered or empirical, start out assuming at least 16" (41 cm) of masonry between openings, 24" (61 cm) if possible to provide the mason with a fighting chance of placing vertical reinforcing between the lintels.

Other planning issues: With respect to the diaphragm length over width ratio. Earlier we discovered the building required an interior shear wall of some type to break the diaphragm span down to no more than a 2:1 span to depth ratio. Because the structure is 60' (18.3 m) long, assume this wall will be at 30' (9.14 m) so we assume it will be a wood shear wall that is located about mid-length, perhaps as a stair wall. If the architectural plans are too far along to change course, then perhaps we add a steel moment frame to laterally brace the structure. Moment frames of steel can be flexible so check deflection and compare it to the permitted deflection for a masonry bearing wall. Masonry is brittle and deflection compatibility is a major concern during design.

Use the code for determining how much wall is needed in each direction to resist shear forces. The minimum cumulative length of wall in each direction is 0.2 times the long or broad face dimension. This is relatively easy to obtain for our example building, i.e. 0.2 × 60=12' (3.66 m) thus, we only need 12' (3.66 m) of wall along each of the four faces.

This as background, the author believes that some projects may be designed using the empirical method but that it's best used during the preliminary stages of design but that more rigorous methods should be utilized for final design.

Computer design packages have come a long way since ASA A41.1 was published in 1944. Consider purchasing a design package if you are going into this type of work in earnest. Also, download a copy of TM 5-809-3 which is the Army/Navy/Air Force tech manual for masonry and available for free from the US Army Corps of Engineers website. It contains a wealth of common sense information, much of it timeless, but be careful because the one third increase in allowable stress for combinations that include wind or seismic loads no longer applies. Instead, the current code combinations include load (or probability) factors on the demand side of the equation and the supply side is to remain unfactored.

Last, make sure the owner engages the services of a third-party inspector. Inspection is required by both IBC and the Specification for Masonry Structures TMS 602/ACI 530.1/ASCE 6. Consult the code for the appropriate level of inspection. Make sure the reference is accurate, IBC versus TMS 602/ACI 530.1/ASCE 6 because terminology between them is not perfectly aligned. Also, both rely on the building classifications found in ASCE 7.

Avoid prism testing because of its cost and difficulty. Mortar testing is also discouraged because of its variability and because it can't be used in acceptance of the final masonry assembly. In lieu of these testing costs, review the material submittals including manufacturer's test results to determine whether the assembly will be in compliance with the required strength. Encourage full-time on-site inspection by a knowledgeable and qualified inspection firm to verify reinforcing steel is placed correctly including size, location, lap length, etc. and to monitor mortar mixing, retempering, and grout placement are all in accordance with the drawings and specifications. There are horror stories the author is aware of. One, a school building that was closed for over a year to remediate voids due to improper grouting.

In closing, the design of structures is not for the faint of heart, the uninformed, or the careless. Always check your answers using alternate methods, either by using software or by hand. Always check computer generated answers assuming they are wrong until proven right; mistakes in the problem definition will produce mistakes in the results. Garbage in = garbage out.

NOTES

1. *Oxford English Dictionary.* 2nd ed. 20 vols (Oxford: Oxford University Press, 1989).
2. "The Code of Hammurabi," Constitution Society, accessed 11/2017. www.constitution.org/ime/hammurabi.htm
3. "Thatching in the City of London," *Thatching Info,* accessed 11/2017. http://thatchinginfo.com/thatching-in-the-city-of-london/
4. "History of Building Regulations," *Building History,* accessed 11/2017. www.buildinghistory.org/regulations.shtml
5. B. Fletcher, *London Building Acts* (London: BT Batsford, 1914).
6. ICBO, *Uniform Building Code* (International Conference of Building Officials, Long Beach California, 1928).
7. TMS, *Building Code Requirements for Masonry Structures,* TMS 402–13. (Longmont, CO: The Masonry Society, 2013).
8. OSHA, *Requirements for Masonry Construction,* 1926.706 (Washington DC: Occupational Safety and Health Administration, 2017).
9. MSJC, *Building Code Requirements for Masonry Structures* (Longmont, CO: The Masonry Society, 2013.

Masonry Bending

Chapter 4

Paul W. McMullin

4.1 Stability
4.2 Capacity
4.3 Demand versus Capacity
4.4 Deflection
4.5 Detailing Considerations
4.6 Lintel Design Example
4.7 Where We Go from Here

Masonry resists bending loads by working in harmony with reinforcing steel. The masonry carries compression, while the steel carries tension. Because the force **resultants** are offset from each other, they create a **couple** that resists bending **moments**.

The most common bending members in masonry construction are walls and lintels, such as that in Figure 4.1. Walls resist out-of-plane bending from wind, seismic, or blast loads. They also resist in-plane shear forces, also due to wind and seismic forces. Lintels are beams that are integral with the wall. They can be made of reinforced masonry, concrete, or steel, illustrated in Figure 4.2. We will focus our efforts on wall and lintel design in this chapter.

4.1 STABILITY

Reinforced masonry lintels with large spans and depths can become unstable and roll over in the middle. Known as lateral torsional buckling, the compressive force in the top of the lintel can become great enough that it causes **buckling**, forcing the middle outward. This is illustrated in Figure 4.3 for a very slender wood beam. TMS 402 requires that the compression face be braced as follows:

$$s = \min \left| \begin{array}{l} 32b \\ 120 \dfrac{b^2}{d} \end{array} \right. \quad (4.1)$$

Figure 4.1 Masonry wall and lintel in an industrial building

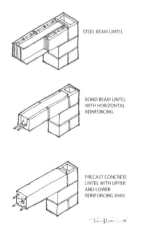

Figure 4.2 Masonry lintel types, courtesy Teran Mitchell

Figure 4.3 Lateral torsional buckling in wood beam

where:

- b = member width, in (mm)
- d = distance from furthest compression edge to centroid of the reinforcing steel, in (mm)

Based on these limits, Table 4.1 provides brace spacing requirements for a variety of masonry lintel widths and depths. Where bracing is required, roof and floor diaphragms are very effective in bracing the compression regions of the beam. Where this is not possible, kickers can be used.

4.2 CAPACITY

Engineers regularly make assumptions regarding support conditions, loading, stress/strain relationships, and internal equilibrium to simplify calculations. Although assumptions are dangerous when inaccurate,

Table 4.1 Lintel lateral bracing requirements

Imperial Depth (in)	*Lateral Brace Spacing (ft)* Nominal Width (in)			
	8	10	12	16
8	20.3	25.7	31.0	41.7
16	20.3	25.7	31.0	41.7
24	20.3	25.7	31.0	41.7
32	18.2	25.7	31.0	41.7
40	14.5	23.2	31.0	41.7
48	12.1	19.3	28.2	41.7
56	10.4	16.5	24.1	41.7
64	9.08	14.5	21.1	38.1
72	8.08	12.9	18.8	33.9
80	7.27	11.6	16.9	30.5
88	6.61	10.5	15.4	27.7
Metric Depth (mm)	*Lateral Brace Spacing (m)* Nominal Width (mm)			
	203	*254*	*305*	*406*
203	6.19	7.82	9.46	12.7
406	6.19	7.82	9.46	12.7
609	6.19	7.82	9.46	12.7
812	5.53	7.82	9.46	12.7
1,015	4.43	7.07	9.46	12.7
1,218	3.69	5.89	8.60	12.7
1,421	3.16	5.05	7.37	12.7
1,624	2.77	4.42	6.45	11.6
1,827	2.46	3.93	5.73	10.3
2,030	2.21	3.53	5.16	9.29
2,233	2.01	3.21	4.69	8.45

they make the problem manageable. The following five assumptions govern the analysis and design of masonry elements subjected to **flexure** and can be found in TMS 402.[1]

Assumption #1—Strain in the masonry and steel reinforcing is proportional to the distance from the **neutral axis** (line of no strain).

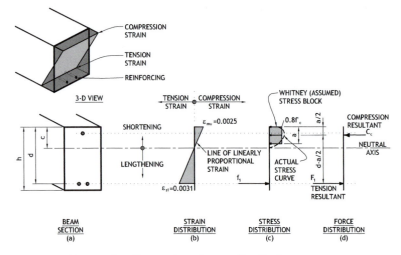

Figure 4.4 Masonry bending strain and stress distribution

The amount of lengthening or shortening (strain) in the beam makes a straight line between the edge of the beam that is in compression and reinforcing steel that is in tension, shown in Figure 4.4b.

Assumption #2—The maximum useable strain in concrete masonry ε_{cu} = 0.0025 (or 0.25%), and brick masonry crushes at a strain ε_{bu} of 0.0035. The masonry begins to crush and rapidly loose strength when strained beyond this point.

Assumption #3—Stress in the reinforcing steel equals the yield strength f_y, if the steel has reached its yield strain ε_y. Otherwise, it is equal to $E_s \varepsilon_s$. As strain increases, there is a proportional increase in stress until the steel yields. Thereafter, the stress remains at the yield level and does not increase even though strain does (recall Figure 2.26). This is advantageous—and at the heart of many aspects of concrete bending design—because it allows the steel to stretch significantly without breaking.

Assumption #4—The tensile strength of the masonry can be ignored. Because it is much weaker in tension than reinforcing steel, it contributes little to tension strength.

Assumption #5—The compression **stress block** can be defined as follows:

- Stress equals $0.80f'_m$.
- The stress is uniform across the width b of the beam.
- The height of the stress block a equals $0.8c$, where c is the distance to the neutral axis from the extreme compression fiber.

Figure 4.4c provides a graphical representation of assumption #5.

Utilizing these assumptions, we can determine the tension force in the reinforcing, and resultant in the compression block (see Figure 4.4d). We then find the nominal flexural capacity M_n by summing the moments these forces create, as they are offset from each other—discussed further in section 4.2.2.

4.2.1 Reinforcement Locations

Understanding how beams deform helps us know where to put reinforcing steel in masonry beams. Simple span beams with downward load experience tension in the bottom and compression in the top at the mid-span. Figure 4.5a shows a simply supported foam beam with a **point load** in the middle. Notice how the circles on the bottom middle stretch—indicating tension—and shorten at the top—indicating compression. Taking this further, Figure 4.5b shows a multi-span beam with point loads. The middle deformation is the same as a simply supported beam, but over the middle supports, where the beam is continuous, the tension and compression change places—tension on top, compression on bottom. If this beam were **fixed** at the ends, it would have the same deformation as the middle support.

To simplify analysis, we often assume masonry lintels are simply supported beams. This creates tension in the bottom and compression in the top. For tension, we place reinforcing steel in the bottom of the lintel, like that in Figure 4.6. We typically match the tension steel in the compression zone of the lintel for additional integrity and to hang shear bars.

4.2.2 Flexural Strength

Applying the five assumptions, and knowing where we need primary reinforcing, we choose the reinforcing bar area and determine the flexural strength at key locations along the beam.

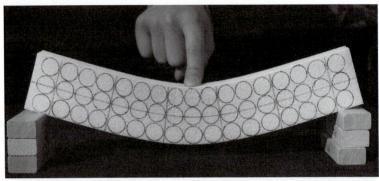

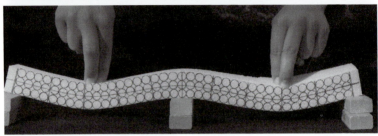

Figure 4.5 Deformed shape of (a) simply supported and (b) multi-span beams

To calculate bending strength, we must determine the tension force in the reinforcing steel T (k or kN) and compressive block resultant C (k or kN)—shown in Figure 4.4d—using the following equations:

$$T = A_s f_y \qquad (4.2)$$

where:

A_s = **cross-sectional area** of the reinforcing, in² (mm²)
f_y = reinforcing yield strength, typically 60 k/in² (420 MN/m²)

$$C = 0.80 f'_m ba \qquad (4.3)$$

where:

C = compressive force in the masonry, k (kN)
f'_m = compressive strength of the masonry, from 1,700 to 3,000 lb/in² (11.72 to 20.68 MN/m²)
b = beam width, in (mm)
a = compression block depth, in (mm)

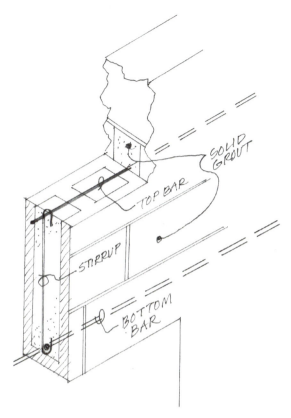

Figure 4.6 Reinforcing locations in a masonry lintel

Before doing any calculations, we specify the masonry strength, assume beam dimensions, and the amount of reinforcing steel required. The Initial Lintel Sizing box provides a simple method for a first pass for the reinforcing steel required.

Once we have compressive strength, dimensions, and reinforcing steel, the only unknown in equations (4.2) and (4.3) is a. We solve for a by setting the force in the reinforcing steel equal to the concrete compressive force.

$$C = T \qquad 0.8f'_m ba = A_s f_y \qquad (4.4)$$

Rearranging we get:

$$a = \frac{A_s f_y}{0.8 f'_m b} \tag{4.5}$$

Summing moments about the compression block resultant (shown in Figure 4.4d), we derive the following expression for nominal bending capacity:

$$M_n = T\left(d - \frac{a}{2}\right) \tag{4.6}$$

where

d = the distance from the top of the compression block to the centerline of the tension reinforcing steel, in (mm)

Finally, we multiply the **nominal strength** by the strength reduction factor ϕ of 0.9 to get the design capacity ϕM_n.

BOX 4.1 INITIAL LINTEL SIZING

For initial lintel depth d_{est}, we can use the following relationship, which is also used in Table 2.1.

$$d_{est} = \frac{l}{5} \tag{4.7}$$

where

l = lintel span, in (mm)

Using this limit also exempts us from checking deflection under normal loading.

For the initial area of steel reinforcing A_s, we can take something between 0.002 and 0.006 times the gross section area. Use the lower value for lightly loaded lintels, and the higher for heavier loads.

Table 4.2 provides a good starting point to estimate lintel depth and required reinforcing for various spans.

Table 4.2 Initial lintel sizing aid

Span	Height	Masonry Width (in)					
l	h	8		10		12	
(ft)	(in)	QTY	Bar	QTY	Bar	QTY	Bar
4	8	(1)	#4	(1)	#4	(1)	#4
6	16	(1)	#4	(1)	#4	(1)	#5
8	16	(1)	#5	(1)	#4	(1)	#5
10	16	(1)	#5	(1)	#5	(1)	#5
14	24	(1)	#6	(1)	#6	(1)	#6
18	32	(1)	#7	(1)	#7	(1)	#6
22	40	(1)	#8	(1)	#8	(1)	#7
26	48	(1)	#8	(1)	#8	(1)	#8
30	56	(2)	#6	(1)	#9	(1)	#9
34	64	(2)	#7	(2)	#7	(2)	#7
38	72	(2)	#7	(2)	#7	(2)	#7

Span	Height	Masonry Width (mm)					
l	h	203		254		305	
(m)	(mm)	QTY	Bar	QTY	Bar	QTY	Bar
1.22	203	(1)	M13	(1)	M13	(1)	M13
1.83	406	(1)	M13	(1)	M13	(1)	M16
2.44	406	(1)	M16	(1)	M13	(1)	M16
3.05	406	(1)	M16	(1)	M16	(1)	M16
4.27	610	(1)	M19	(1)	M19	(1)	M19
5.49	813	(1)	M22	(1)	M22	(1)	M19
6.71	1,016	(1)	M25	(1)	M25	(1)	M22
7.92	1,219	(1)	M25	(1)	M25	(1)	M25
9.14	1,422	(2)	M19	(1)	M29	(1)	M29
10.4	1,626	(2)	M22	(2)	M22	(2)	M22
11.6	1,829	(2)	M22	(2)	M22	(2)	M22

Notes: 1) This table is for preliminary design only. All sizes and reinforcement must be determined based on actual conditions.
2) Based on 50 lb/ft² (2.39 kN/m²) live load, 20 ft (6.1 m) span, maximum reinforcing area and diameter, and minimum reinforcing.

4.3 DEMAND VERSUS CAPACITY

Maintaining an appropriate margin of safety is fundamental to structural engineering. We do this through safety factors—load factors in **load combinations**, and strength reduction factors ϕ applied to the materials. We also increase safety by controlling what portion fails first. This provides energy absorption and warning signals of a problem.

The strength reduction factor for masonry in bending is 0.9. With this we calculate the design bending capacity ϕM_n. The lintel has adequate capacity when the following expression is met.

$$M_u \leq \phi M_n \tag{4.8}$$

where:

M_u = is the flexural demand (moment)

If this expression is not met, we select a larger beam or more reinforcing, and recalculate its strength.

Bending moments for a single span beam are as follows:

Pinned Ends   Fixed Ends

$$M_u = \frac{w_u l^2}{8} \qquad\qquad M_u = \frac{w_u l^2}{11} \tag{4.9}$$

where:

w_u = factored uniform **distributed load**, k/ft (kN/m)
l = **span length**, ft (m)

We also limit the tension reinforcing steel so it yields well before the masonry crushes, as discussed in Section 2.4.3. Using equation 2.7, and α=1.5 for lintels, we can calculate the maximum reinforcing steel A_{max}. Table 4.3 provides these values reinforcing for various combinations of bd.

4.4 DEFLECTION

Deflection can lead to masonry cracking and damage to non-structural elements. For reinforced masonry the typical code limits are applicable, though a more stringent criterion such as $l/360$ or $l/480$ is prudent and easily achievable. For members supporting unreinforced masonry the code limit is $l/600$ for total load. This limit also applies to masonry with only vertical reinforcing. Table 4.4 provides common deflection limits for varying spans and criteria. When a beam span is less than $8d$, deflection does not need to be checked, as it will inherently meet the $l/600$ criteria.

When checking deflection, we use the cracked moment of inertia I_{cr}, as the entire cross-section will not be providing stiffness. We can approximate cracked moment of inertia as 50% of the gross (using the equations in Table 2.15). Use the following equations to calculate

Table 4.3 Maximum reinforcing area in masonry lintels

	Maximum Reinforcing Area A_{max}, in^2						
bd	Masonry Compressive Strength (lb/in^2)						
(in^2)	1,700	1,900	2,000	2,250	2,500	2,750	3,000
ρ_{max}	0.00809	0.00904	0.00952	0.01071	0.01190	0.01309	0.01428
46	0.372	0.416	0.438	0.493	0.547	0.602	0.657
64	0.518	0.579	0.609	0.685	0.761	0.838	0.914
80	0.647	0.723	0.761	0.857	0.952	1.05	1.14
96	0.777	0.868	0.914	1.03	1.14	1.26	1.37
120	0.971	1.09	1.14	1.28	1.43	1.57	1.71
144	1.16	1.30	1.37	1.54	1.71	1.88	2.06
160	1.29	1.45	1.52	1.71	1.90	2.09	2.28
192	1.55	1.74	1.83	2.06	2.28	2.51	2.74
240	1.94	2.17	2.28	2.57	2.86	3.14	3.43
288	2.33	2.60	2.74	3.08	3.43	3.77	4.11
320	2.59	2.89	3.05	3.43	3.81	4.19	4.57
384	3.11	3.47	3.65	4.11	4.57	5.03	5.48
448	3.63	4.05	4.26	4.80	5.33	5.86	6.40
480	3.88	4.34	4.57	5.14	5.71	6.28	6.85
560	4.53	5.06	5.33	6.00	6.66	7.33	8.00
672	5.44	6.08	6.40	7.20	8.00	8.79	9.59
768	6.21	6.94	7.31	8.22	9.14	10.1	11.0
864	6.99	7.81	8.22	9.25	10.3	11.3	12.3

(*continued*)

Table 4.3 continued

	Maximum Reinforcing Area A_{max}, mm²						
bd	Masonry Compressive Strength (MN/m²)						
(mm²)	11.7	13.1	13.8	15.5	17.2	19.0	20.7
ρ_{max}	0.00809	0.00904	0.00952	0.01071	0.01190	0.01309	0.01428
29,677	240	268	282	318	353	388	424
41,290	334	373	393	442	491	540	589
51,613	418	467	491	553	614	675	737
61,935	501	560	589	663	737	811	884
77,419	626	700	737	829	921	1,013	1,105
92,903	752	840	884	995	1,105	1,216	1,326
103,226	835	933	982	1,105	1,228	1,351	1,474
123,871	1,002	1,120	1,179	1,326	1,474	1,621	1,768
154,838	1,253	1,400	1,474	1,658	1,842	2,026	2,211
185,806	1,503	1,680	1,769	1,990	2,211	2,432	2,653
206,451	1,670	1,867	1,965	2,211	2,456	2,702	2,947
247,741	2,004	2,240	2,358	2,653	2,947	3,242	3,537
289,032	2,338	2,613	2,751	3,095	3,439	3,783	4,126
309,677	2,505	2,800	2,947	3,316	3,684	4,053	4,421
361,290	2,923	3,267	3,439	3,869	4,298	4,728	5,158
433,548	3,508	3,920	4,126	4,642	5,158	5,674	6,190
495,483	4,009	4,480	4,716	5,305	5,895	6,484	7,074
557,418	4,510	5,040	5,305	5,969	6,632	7,295	7,958

1) Values assume no axial load.
2) Values for concrete masonry units.
3) Assumes fully grouted cells, or that compression remains in face shells.

deflection for single span beams, with a uniform distributed load. Remember to follow your units through.

Pinned Ends Fixed Ends

$$\delta = \frac{5wl^4}{384 EI_{cr}} \qquad \delta = \frac{wl^4}{384 EI_{cr}} \tag{4.10}$$

where:

    w = uniform distributed load, k/ft (kN/m)
    l = span length, ft (m)
    E = modulus of elasticity, k/in² (MN/m²)
    I_{cr} = moment of inertia, in⁴ (mm⁴)

Table 4.4 Deflection limits for varying lengths and criteria

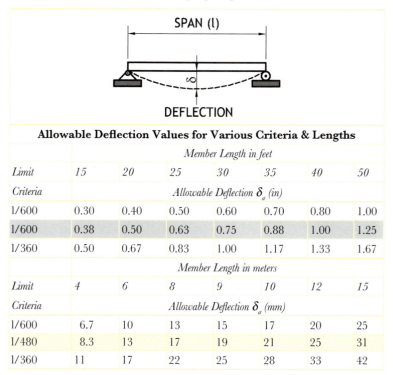

Allowable Deflection Values for Various Criteria & Lengths							
	Member Length in feet						
Limit	*15*	*20*	*25*	*30*	*35*	*40*	*50*
Criteria	*Allowable Deflection δ_a (in)*						
l/600	0.30	0.40	0.50	0.60	0.70	0.80	1.00
l/600	0.38	0.50	0.63	0.75	0.88	1.00	1.25
l/360	0.50	0.67	0.83	1.00	1.17	1.33	1.67
	Member Length in meters						
Limit	*4*	*6*	*8*	*9*	*10*	*12*	*15*
Criteria	*Allowable Deflection δ_a (mm)*						
l/600	6.7	10	13	15	17	20	25
l/480	8.3	13	17	19	21	25	31
l/360	11	17	22	25	28	33	42

4.5 DETAILING CONSIDERATIONS

Based on experience, and to address other considerations beside strength and deflection, the code requires consideration of the way the lintel is configured and has limits on the reinforcing. Let's look at these.

4.5.1 Lintel Configuration

A typical lintel configuration is shown in Figure 4.7. It shows limits on reinforcing size, bar spacing, grout clearance, and cover requirements. It can be used with the tables in this chapter and Chapter 2 to properly configure a lintel.

4.5.2 Minimum Reinforcing

Masonry design does not have an explicit requirement for minimum reinforcing steel. Rather, it requires that the nominal capacity M_n be greater than 1.3 times the cracking moment M_{cr}. This ensures there is enough reinforcing steel after the masonry cracks. The cracking moment is given as follows:

$$M_{cr} = f_r S \tag{4.11}$$

where:

f_r = is the modulus of rupture, lb/in² (kN/m²), from Table 2.6
S = section modulus of uncracked member, in³ (mm³)

Remember to watch your units.

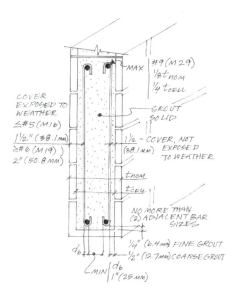

Figure 4.7 Typical lintel reinforcing requirements

BOX 4.2 DESIGN STEPS

These steps will help guide you through the design process. As you apply them to other design tasks, think through how things might change. This can be aided by drawing out the problem, reading additional code sections or textbooks, and exploring the Where Do We Go from Here section.

1. Draw the structural layout, include span dimensions and **tributary width**
2. Determine Loads—Unit, load combinations generating a line load, member bending moment
3. Material Parameters—Choose masonry compressive strength and reinforcing strength
4. Estimate initial masonry size and reinforcing
5. Determine the bending strength
6. Check maximum and minimum reinforcing steel
7. Check deflection
8. Summarize the results

4.6 DESIGN EXAMPLE

Step 1: Structural Layout

Our example structure is a small maintenance shop. It is 40'-0" (12.19 m) across and 60'-0' (18.29 m) long. It has two openings for equipment to enter, and two side doors, shown in Figure 4.8. We will design the 16'-0" (4.88 m) lintel for flexure in this example. Perhaps you can design the 10'–0" (3.05 m) lintel for fun later.

From the general layout, we draw the loading conditions for our lintel in Figure 4.9. Because the lintel is so close to the top of the wall, we cannot assume arching action. Thus, the full weight of the wall above the opening and full roof load will be carried by our lintel. No big deal. From our sketches we get the following geometric data:

$l = 16 ft$	$l = 4.88 m$
$l_t = 19.33 ft$	$l_t = 5.89 m$
$h_a = 5.0 ft$	$h_a = 1.52 m$

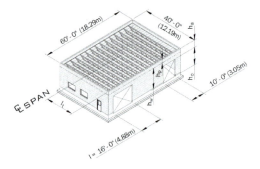

Figure 4.8 Example building configuration

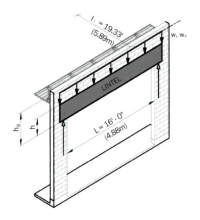

Figure 4.9 Example lintel layout

Because the roof is near the top of the lintel, we can say it is fully braced for lateral torsional buckling.

Step 2: Determine the Loads

Our roof dead and **snow loads** are as follows:

$$D_R = 20\frac{lb}{ft^2}$$

$$S = 50\frac{lb}{ft^2}$$

$$D_R = 0.96\frac{kN}{m^2}$$

$$S = 2.39\frac{kN}{m^2}$$

We will also need to know the unit weight of the masonry lintel. We will assume an 8 in (203 mm) thick wall. Our masonry is lightweight, and we have a solid grouted lintel. Thus, from Table 2.3, we get:

$$D_W = 78 \frac{lb}{ft^2} \qquad\qquad D_W = 3.73 \frac{kN}{m^2}$$

We now calculate the factored line load on the lintel. Because snow load controls, we use the snow favored load combination. We will need an unfactored line load for deflection, and a factored one for strength.

$$w = (D_R + S)l_t + D_W h_a$$

$$= \left[\left(20\frac{lb}{ft^2} + 50\frac{lb}{ft^2}\right)19.3ft + 78\frac{lb}{ft^2}(5.0ft)\right]\frac{1k}{1,000lb}$$

$$= 1.74 \frac{k}{ft}$$

$$= \left(0.96\frac{kN}{m^2} + 2.39\frac{kN}{m^2}\right)5.89m + 3.73\frac{kN}{m^2}(1.52m)$$

$$= 25.4 \frac{kN}{m}$$

$$w_u = (1.2D_R + 1.6S)l_t + 1.2D_W h_a$$

$$= \left[\left(1.2\left(20\frac{lb}{ft^2}\right) + 1.6\left(50\frac{lb}{ft^2}\right)\right)19.3ft + 1.2\left(78\frac{lb}{ft^2}\right)(5.0ft)\right]\frac{1k}{1,000lb}$$

$$= 2.48 \frac{k}{ft}$$

$$= \left(1.2\left(0.96\frac{kN}{m^2}\right) + 1.6\left(2.39\frac{kN}{m^2}\right)\right)5.89m + 1.2\left(3.73\frac{kN}{m^2}\right)(1.52m)$$

$$= 36.1 \frac{kN}{m}$$

We now find the bending moment in the lintel. We will assume it is a simply supported lintel. While the ends are realistically fixed, there are times lintels are near the ends of walls that provide little fixity.

$$M_u = \frac{w_u l^2}{8}$$

$$= \frac{2.48 k/ft (16ft)^2}{8} \qquad\qquad = \frac{36.1 kN/m (4.88m)^2}{8}$$

$$= 79.4 k-ft \qquad\qquad\qquad\quad = 108 kN-m$$

Step 3: Masonry Parameters

We now need to determine our masonry and reinforcing strengths, and masonry modulus of elasticity. From Table 2.5 we get the masonry compressive strength.

$$f'_m = 1,700 \frac{lb}{in^2} \qquad\qquad f'_m = 11.7 \frac{MN}{m^2}$$

From this, we can calculate the modulus of elasticity:

$$E_m = 900 f'_m$$
$$= 900 \left(1.7 \frac{k}{in^2}\right) \qquad\qquad = 900 \left(11.7 \frac{MN}{m^2}\right)$$
$$= 1,530 \frac{k}{in^2} \qquad\qquad = 10,530 \frac{MN}{m^2}$$

The reinforcing steel yield strength is:

$$f_y = 60 \frac{k}{in^2} \qquad\qquad f_y = 420 \frac{MN}{m^2}$$

Step 4: Initial Size

Let's now estimate our lintel size and reinforcing. From Table 4.2 we get the following initial size and reinforcing for a 18'-0 (5.49 m) span:

$h = 32 in$ $\qquad\qquad h = 813 mm$

One #7 (M22) bar, giving us an area of:

$A_s = 0.60 in^2$ $\qquad\qquad A_s = 387 mm^2$

Looking at Table 2.9 we see the maximum bar size is a #8 (M25), so we know we are OK.

Step 5: Bending Strength

We now have what we need to calculate and check the strength. Looking at the cross-section in Figure 4.10, we see the distance between the bottom of the lintel and center of reinforcing is:

$d' = 2.063 in$ $\qquad\qquad d' = 53 mm$

From this we can calculate the distance from the reinforcing to top of lintel.

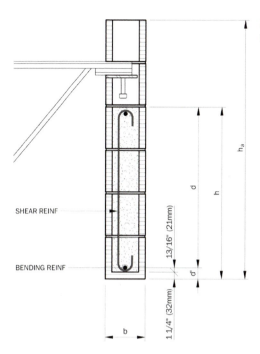

Figure 4.10 Example lintel cross-section

$$d = h - d'$$
$$= 32in - 2.063in \qquad\qquad = 813mm - 53mm$$
$$= 29.9in \qquad\qquad\qquad\;\; = 761mm$$

We will also need the masonry width b, which is 3/8 in (9.5 mm) less than the nominal width. Thus,

$$b = 8in - \frac{3}{8}in \qquad\qquad b = 203mm - 9.5mm$$
$$= 7.625in \qquad\qquad\quad\;\; = 194mm$$

We next find the tension force in the reinforcing, and compression block depth.

$$T = A_s f_y$$
$$= 0.60in^2\left(60\frac{k}{in^2}\right) \qquad = 387mm^2\left(420\frac{MN}{m^2}\right)\left(\frac{1m}{1,000mm}\right)^2\frac{1,000kN}{1MN}$$
$$= 36k \qquad\qquad\qquad\qquad\; = 163kN$$

Bending

$$a = \frac{T}{0.8f'_m b}$$

$$= \frac{36k}{0.8\left(1.7\dfrac{k}{in^2}\right)7.625in} \qquad = \frac{163kN}{0.8\left(11,700\dfrac{kN}{m^2}\right)194mm\left(\dfrac{1m}{1,000mm}\right)^2}$$

$$= 3.47in \qquad\qquad\qquad\qquad = 89.8mm$$

The nominal bending capacity follows as:

$$M_n = T\left(d - \frac{a}{2}\right)$$

$$= 36k\left(29.94in - \frac{3.47in}{2}\right)\frac{1k}{12in} \qquad = 163kN\left(0.761m - \frac{0.090}{2}\right)$$

$$= 84.6k\text{-}ft \qquad\qquad\qquad\qquad = 117kN\text{-}m$$

Multiplying by a strength reduction factor f of 0.9, we get:

$$\phi M_n = 0.9M$$

$$= 0.9(84.6k\text{-}ft) \qquad\qquad = 0.9(117kN\text{-}m)$$

$$= 76.1k\text{-}ft \qquad\qquad\qquad = 105kN\text{-}m$$

This is slightly less than our demand, so we know our lintel doesn't work. This drives home the point that preliminary tables are good for just that: the first pass of design. Because the table we used was derived based on a 10'-0 (3.05 m) tributary width, and ours is 20'-0 (6.1 m) it isn't a surprise that it didn't work out. Let's try one #8 (M25) bar, and see what that does for us.

$$A_s = 0.79in^2 \qquad\qquad A_s = 510mm^2$$

$$T = A_s f_y$$

$$= 0.79in^2\left(60\frac{k}{in^2}\right) \qquad = 510mm^2\left(420\frac{MN}{m^2}\right)\left(\frac{1m}{1,000mm}\right)^2\frac{1,000kN}{1MN}$$

$$= 47.4k \qquad\qquad\qquad = 214kN$$

$$a = \frac{T}{0.8f'_m b}$$

$$= \frac{47.4k}{0.8\left(1.7\dfrac{k}{in^2}\right)7.625in} \qquad = \frac{214kN}{0.8\left(11,700\dfrac{kN}{m^2}\right)194mm\left(\dfrac{1m}{1,000mm}\right)^2}$$

$$= 4.57in \qquad\qquad\qquad\qquad = 118mm$$

$$M_n = T\left(d - \frac{a}{2}\right)$$

$$= 47.4k\left(29.94in - \frac{4.57in}{2}\right)\frac{1k}{12in} \qquad = 214kN\left(0.761m - \frac{0.118mm}{2}\right)$$

$$= 109k-ft \qquad\qquad\qquad\qquad\qquad = 150kN-m$$

$$\phi M_n = 0.9(109k-ft) \qquad\qquad\qquad \phi M_n = 0.9(150kN-m)$$

$$= 98.1k-ft \qquad\qquad\qquad\qquad\qquad = 135kN-m$$

This is larger than the demand M_u, so we are OK for strength.

Step 6: Check Maximum and Minimum Reinforcing

Now, we need to check the maximum and minimum reinforcing steel. It can be helpful to do this before calculating strength, but it's certainly OK to do this at the end. To determine the maximum steel, we need bd, and use Table 4.3.

$bd = 7.625in\,(29.94in)$ $\qquad\qquad\qquad bd = 194mm\,(761mm)$
$\quad = 228in^2$ $\qquad\qquad\qquad\qquad\qquad\quad = 1.48 \times 10^5\,mm^2$

This gives a maximum steel area of:

$A_{max} = 1.55in^2$ $\qquad\qquad\qquad\qquad A_{max} = 1,000mm^2$

To check the minimum steel, we first calculate the section modulus and find the modulus of rupture.

$$S_x = \frac{1}{6}bd^2$$

$$= \frac{1}{6}(7.625in)(29.94in)^2 \qquad\qquad = \frac{1}{6}(194mm)(761mm)^2$$

$$= 1{,}139\,in^3 \qquad\qquad\qquad\qquad\quad = 18.7 \times 10^6\,mm^3$$

From Table 2.6 we get:

$$f'_r = 267\,\frac{lb}{in^2} \qquad\qquad\qquad\qquad f'_r = 1{,}839\,\frac{kN}{m^2}$$

Finding the cracking moment M_{cr}, we get:

$$M_{cr} = f'_r S_x$$

$$= 0.267 \frac{k}{in^2} 1{,}139 in^3 \frac{1ft}{12in} \qquad = 1{,}839 \frac{kN}{m^2} 18.7 \times 10^6 mm^3 \frac{1^3 m^3}{1{,}000^3 mm^3}$$

$$= 25.3 k\text{-}ft \qquad\qquad\qquad\qquad = 34.4 kN\text{-}m$$

Because our nominal capacity M_n is greater than 1.3 M_{cr}, we know we have enough reinforcing steel.

Step 7: Check Deflection

Even though we don't need to check deflection because our lintel is deeper than l/5, let's do it anyway. First, we calculate the cracked moment of inertia I_{cr}.

$$I_{cr} = 0.5 \left(\frac{1}{12} bd^3 \right)$$

$$= 0.5 \left(\frac{1}{12} (7.625 in)(29.94 in)^3 \right) \qquad = 0.5 \left(\frac{1}{12} (194 mm)(761 mm)^3 \right)$$

$$= 8{,}527 in^4 \qquad\qquad\qquad\qquad\qquad = 3.56 \times 10^9 mm^4$$

Then deflection,

$$\delta = \frac{5wl^4}{384 E_m I_{cr}}$$

$$= \frac{5 \left(1.74 \frac{k}{ft} \right) (16 ft)^4 \left(\frac{12 in}{1 ft} \right)^3}{384 \left(1{,}530 \frac{k}{in^2} \right) 8{,}527 in^4}$$

$$= 0.197 in$$

$$= \frac{5 \left(0.0253 \frac{MN}{m} \right) (4.88 m)^4}{384 \left(10{,}530 \frac{MN}{m^2} \right) 3.56 \times 10^9 mm^4 \left(\frac{1 m}{1{,}000 mm} \right)^5}$$

$$= 4.98 mm$$

Using a stringent criterion of l/600, we can calculate our allowable deflection.

$$\delta_a = \frac{l}{600}$$

$$= \frac{16ft}{600}\left(\frac{12in}{1ft}\right) \qquad = \frac{4.88m}{600}\left(\frac{1,000mm}{1m}\right)$$

$$= 0.32in \qquad\qquad\qquad = 8.13mm$$

The allowable is greater than the actual deflection, so our design works.

Step 8: Summarize Results

In summary, we have an 8 in (200 mm) wide by 32 in (810 mm) deep lintel. It has one #8 (M25) bar in the top and bottom and is fully grouted.

And just like that, we have our bending design!

4.7 WHERE WE GO FROM HERE

This chapter presents the basics of reinforced masonry beam design. With a sound understanding of these principles, you have the basic tools to design masonry flexural members. We will extend these principles to wall and column design. Shear is another aspect of beam design and will be considered in Chapter 5.

NOTE

1. TMS, *Building Code Requirements for Masonry Structures*, TMS 402–13 (Longmont, CO: The Masonry Society, 2013).

Masonry Shear

Chapter 5

Paul W. McMullin

5.1 Stability
5.2 Capacity
5.3 Demand versus Capacity
5.4 Deflection
5.5 Detailing
5.6 Design Example
5.7 Where We Go from Here

Shear is fundamental to how bending members resist load. It holds the outer layers together, causing them to act as one. This chapter will focus on how we size lintels for shear stress. Chapter 7 covers shear wall design to resist wind and seismic forces.

Let's take a moment and conceptually understand the fundamentals of shear behavior. You are likely familiar with the action scissors make when cutting paper or fabric. The blades are perpendicular to the material, going in opposite directions. This creates a tearing of the material like that shown in Figure 5.1. In beam shear, the action is similar, but the movement of material is parallel to the length of the beam. The top portion moves relative to the bottom, illustrated in Figure 5.2.

Shear strength is fundamentally tied to bending strength and stiffness. If we take a stack of paper and lay it across two supports, it sags (Figure 5.3a), unable to carry even its own load. If we glue each strip of paper together, we get a beam with enough strength and stiffness to carry a reasonable load, as shown in Figure 5.3b. And so, it is with masonry beams. The lengthwise fibers provide bending strength, but it is the material in the middle that holds them together.

5.1 STABILITY

Shear stability is not a concern in masonry members. The sections are compact enough to not experience shear buckling.

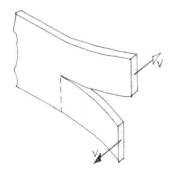

Figure 5.1 Shearing action like scissors

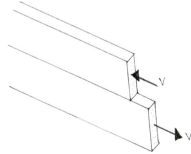

Figure 5.2 Shearing action from bending

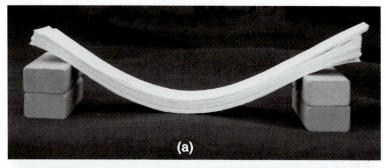

Figure 5.3 Paper beam with layers (a) unglued, and (b) glued

5.2 CAPACITY

Shear strength in masonry members comes from two sources, the masonry and reinforcing steel. In equation form we have:

$$V_n = (V_{nm} + V_{ns})\gamma_g \tag{5.1}$$

where:

V_{nm} = nominal masonry shear strength

V_{ns} = nominal steel shear strength

γ_g = grouted shear wall factor

= 0.75 for partially grouted shear walls, and 1.0 otherwise

The nominal shear capacity is limited to the following values to preclude low deformation shear failures. The limits are as follows:

$$M_U/(V_U d_V) \leq 0.25 \tag{5.2}$$

$$V_{n,\max} \le \left(6A_{nv}\sqrt{f'_m}\right)\gamma_g \qquad V_{n,\max} \le \left(0.498 A_{nv}\sqrt{f'_m}\right)\gamma_g$$

$M_u/(V_u d_v) \ge 1.0$

$$V_{n,\max} \le \left(4A_{nv}\sqrt{f'_m}\right)\gamma_g \qquad V_{n,\max} \le \left(0.332 A_{nv}\sqrt{f'_m}\right)\gamma_g$$

where:

M_u = bending moment, k-ft (kN-m)
V_u = shear demand, k (kN)
d_v = member depth in direction of shear, ft (m)
A_{nv} = net shear area, in² (mm²)

Make sure you keep f'_m in units of lb/in² (MN/m²), as the equation is testing based and requires specific units.

We use linear interpolation between the moment to shear ratios above. We can also use $M_u/(V_u d_v) = 1.0$ for simplicity.

5.2.1 Masonry Shear Capacity V_{nm}

Masonry resists shear through friction in the compression zone, aggregate interlock, and dowel action, illustrated in Figure 5.4. A simplified version of the shear strength contribution from masonry is given as:

(5.3)

$$V_{nm} = 2.25 A_{nv}\sqrt{f'_m} + 0.25 P_u \qquad V_{nm} = 0.187 A_{nv}\sqrt{f'_m} + 0.25 P_u$$

where:

P_u = factored axial load, k (kN), positive for compression

These equations yield units of lb and N, respectively.

Make sure to use the units listed above, since the equation is based on testing.

5.2.2 Reinforcement Shear Capacity V_{ns}

When the shear capacity of the masonry is not sufficient, we provide shear reinforcement. This increases strength and provides higher

Masonry Shear

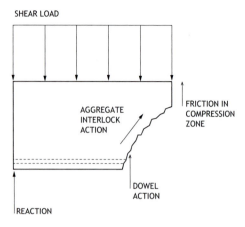

Figure 5.4 Shear mechanisms in masonry

deformation behavior. We determine steel shear strength contribution as follows:

$$V_{ns} = 0.5 \frac{A_v}{s} f_y d_v \tag{5.4}$$

where:

A_v = cross-sectional area of shear reinforcing, in² (mm²)
s = reinforcing spacing, in (mm)
f_y = reinforcing yield strength, k/in² (MN/mm²)

The minimum area of shear reinforcing is:

$$A_{vmin} = 0.0007 b d_v \tag{5.5}$$

See the Detailing section of this chapter for spacing and bar configuration requirements.

Once we have the masonry and steel contributions to shear capacity, we add them together as shown in equation (5.1) and check the limits given in equation (5.2). If our design meets the limits, we multiply the nominal shear capacity V_n by the strength reduction factor ϕ, equal to 0.8, as discussed further in section 5.3.

BOX 5.1 INITIAL LINTEL SHEAR SIZING

This section will help you choose an initial lintel size for shear. For depth, dividing the span by 8 ($l/8$) will give a good starting point. For reinforcing, #4 @ 8 in on center (M13 @ 203 mm) is very common. For heavier loads, two #4 in each vertical cell is effective.

Taking this further, Table 5.1 provides masonry shear strength for various lintel sizes. Similarly, Table 5.2 presents reinforcing shear strength for varying depth to reinforcing spacing (d_v/s) ratios. We can use these to get a feel for lintel size once we have calculated the shear demand V_u.

Table 5.1 Masonry shear strength for varying lintel depths

Imperial					
		\multicolumn{4}{c}{*Masonry Shear Strength* ϕV_{nm} (k)}
		\multicolumn{4}{c}{*Nominal Width b (in)*}
		8	*10*	*12*	*16*
h (in)	8	4.5	5.7	6.9	9.3
	16	9.1	11	14	19
	24	14	17	21	28
	32	18	23	28	37
	40	23	29	35	46
	48	27	34	41	56
	56	32	40	48	65
	64	36	46	55	74
	72	41	51	62	83
	80	45	57	69	93
	88	50	63	76	102

Table 5.1 *continued*

Metric					
		\multicolumn{4}{c}{*Masonry Shear Strength* ϕV_{nm} *(kN)*}
		\multicolumn{4}{c}{*b (mm)*}
		203.2	254	304.8	406.4
h (mm)	203	20	25	31	41
	406	40	51	61	83
	610	60	76	92	124
	813	81	102	123	165
	1,016	101	127	154	206
	1,219	121	153	184	248
	1,422	141	178	215	289
	1,626	161	203	246	330
	1,829	181	229	276	371
	2,032	201	254	307	413
	2,235	222	280	338	454

1) This table is for preliminary design only. All sizes must be determined for the specific loading condition.

Table 5.2 Reinforcing shear strength for varying depth to spacing

Imperial		\multicolumn{7}{c}{*Reinforcing Shear Strength* ϕV_{ns} (k)}
Stirrups	A_v	\multicolumn{7}{c}{d_v/s}
	(in²)	2.0	2.5	3.0	3.5	4.0	5.0	6.0
(1) #3	0.11	5.3	6.6	7.9	9.2	11	13	16
(1) #4	0.20	10	12	14	17	19	24	29
(1) #5	0.31	15	19	22	26	30	37	45
(2) #3	0.22	11	13	16	18	21	26	32
(2) #4	0.40	19	24	29	34	38	48	58
(2) #5	0.62	30	37	45	52	60	74	89

Table 5.2 continued

Metric Stirrups	A_v (mm²)	Reinforcing Shear Strength ϕV_{ns} (kN) d_v/s						
		2.0	2.5	3.0	3.5	4.0	5.0	6.0
(1) M10	71	24	30	36	42	48	60	72
(1) M13	129	43	54	65	76	87	108	130
(1) M16	200	67	84	101	118	134	168	202
(2) M10	142	48	60	72	83	95	119	143
(2) M13	258	87	108	130	152	173	217	260
(2) M16	400	134	168	202	235	269	336	403

1) This table is for preliminary design only. All sizes must be determined for the specific loading condition.

5.3 DEMAND VERSUS CAPACITY

Once we have the shear capacity ϕV_n we compare it to the shear demand V_u. If $\phi V_n \geq V_u$, the member has adequate shear capacity.

The shear force distribution in a beam depends on support conditions and loading. In a single-span, simply supported beam, with a uniform distributed load, the shear stress is zero at the middle and maximum at the ends. A cantilever is the opposite, with maximum force at the supported end. A multi-span beam has maximum shear at the supports. These are illustrated in Figure 5.5.

5.4 DEFLECTION

Shear deflection in masonry lintels is very small compared to bending deflection and is generally ignored. On the contrary, deflection is a key criterion for shear walls resisting lateral loads from earthquakes and wind.

5.5 DETAILING

5.5.1 Shear Reinforcement Requirements

TMS 402[1] has several requirements for shear reinforcing in beams to ensure the reinforcing engages properly, is sufficient in area, and intersects a diagonal crack. These criteria are given below, and in Figure 5.6.

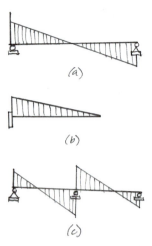

Figure 5.5 Shear force variation at points along (a) simply supported, (b) cantilever, and (c) multi-span beams

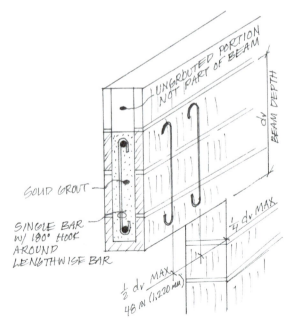

Figure 5.6 Shear reinforcing detailing requirements

- Shear reinforcement is a single bar with a 180 degree hook at the end.
- Shear reinforcement must be hooked around the lengthwise (bending) reinforcing.
- The first shear bar cannot be further than $d_v/4$ from the support.
- The maximum bar spacing must be less than $d_v/4$ or 48 in (1,219 mm).
- Shear reinforcement is useful only when an inclined crack passes through it. Therefore, we don't want to space shear reinforcement too far apart.

From a practical point of view, 180° hooks on bars larger than #5 (M16) get larger than the internal space in concrete block and are best avoided.

BOX 5.2 DESIGN STEPS

These steps will help guide you through the design process. As you apply them to other design tasks, think through how things might change. This can be aided by drawing out the problem, reading additional code sections or textbooks, and exploring the Where Do We Go from Here section.

1. Draw the structural layout, include span dimensions and tributary width
2. Determine Loads—Unit, load combinations generating a line load, member bending moment
3. Material Parameters—Choose masonry compressive strength and reinforcing strength
4. Estimate initial masonry size and reinforcing
5. Determine the shear strength
6. Check maximum and minimum reinforcing steel
7. Summarize the results

5.6 DESIGN EXAMPLE

In this example, we will continue to develop the lintel design we started in Chapter 4. Our lintel spans 16 ft (4.88 m), as shown in Figure 4.8.

Step 1: Structural Layout

From Figure 4.8, and additional information in Figure 5.7, we get the following geometric data.

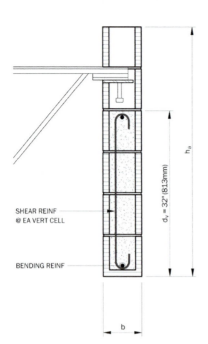

Figure 5.7 Example lintel cross-section

$l = 16 \text{ft}$	$l = 4.88 m$
$d_v = 32 in$	$d_v = 0.81 m$
$b = 7.625 in$	$b = 194 mm$

Step 2: Determine Loads

We calculated the unit and line loads in Chapter 4.

$$w_u = 2.48 \frac{k}{ft} \qquad w_u = 36.2 \frac{kN}{m}$$

Next, we calculate the shear at the end of the lintel, which will be the maximum for our lintel.

$$V_u = \frac{w_u l}{2}$$

$$= \frac{2.48 \frac{k}{ft}(16 ft)}{2} \qquad = \frac{36.2 \frac{kN}{m}(4.88 m)}{2}$$

$$= 19.8 k \qquad = 88.3 kN$$

Step 3: Masonry Parameters

We now need to determine our masonry and reinforcing strengths. From the Chapter 4 example we have:

$$f'_m = 1,700 \frac{\text{lb}}{\text{in}^2}$$

$$f_y = 60 \frac{\text{k}}{\text{in}^2}$$

$$f'_m = 11.7 \frac{MN}{m^2}$$

$$f_y = 420 \frac{MN}{m^2}$$

Step 4: Initial Size

Since we are using the size in Chapter 4, we don't need to worry about this step. Our shear area is therefore:

$A_{nv} = bd_v$

$\quad = 7.625in\,(32in)$ $\qquad\qquad = 194mm\,(810mm)$

$\quad = 244in^2$ $\qquad\qquad\qquad\quad\; = 0.157 \times 10^6 mm^2$

Step 5: Shear Strength

We will now check the contribution of masonry and steel to the shear strength of our lintel. We will also look at the limiting equations on shear capacity.

Step 5a: Masonry Strength

Masonry shear strength is given as:

$$V_{nm} = 2.25 A_{nv} \sqrt{f'_m} \qquad\qquad V_{nm} = 0.187 A_{nv} \sqrt{f'_m}$$

$$V_{nm} = 2.25\,(244in^2)\sqrt{1,700 \frac{\text{lb}}{\text{in}^2}}\,\frac{1k}{1,000lb}$$

$$= 22.6k$$

$$= 0.187\,(0.157 \times 10^6 mm^2)\sqrt{11.7 \frac{MN}{m^2}}\,\frac{1kN}{1,000N}$$

$$= 100.4kN$$

Step 5b: Steel Strength

Let's try (1) #3 at 8 in on center (M10 at 200 mm). The steel area and spacing are:

$A_v = 0.11in^2$ $\qquad\qquad A_v = 71mm^2$

$s = 8in$ $\qquad\qquad\qquad s = 200mm$

Masonry Shear

$$V_{ns} = 0.5 \frac{A_v}{s} f_y d_v$$
$$= 0.5 \left(\frac{0.11 in^2}{8 in} \right) 60 \frac{k}{in^2} (32 in)$$
$$= 13.2 k$$

$$= 0.5 \left(\frac{71 mm^2}{200 mm} \right) 420,000 \frac{kN}{m^2} (810 mm) \left(\frac{1m}{1,000 mm} \right)^2$$
$$= 60.4 kN$$

Step 5c: Check Strength and Reinforcing Limits

Adding the masonry and reinforcing steel contributions together we get:

$V_n = V_{nm} + V_{ns}$
$ = 22.6k + 13.2k \qquad\qquad = 100.4 kN + 60.4 kN$
$ = 35.8k \qquad\qquad\qquad\quad\; = 160.8 kN$

The limiting shear strength is given by:

$$V_{nmax} = 4 A_{nv} \sqrt{f'_m}$$
$$= 4 (244.3 in^2) \sqrt{1,700 \frac{lb}{in^2}} \frac{1k}{1,000 lb}$$
$$= 40.2 k$$

$$V_{nmax} = 0.332 A_{nv} \sqrt{f'_m}$$
$$V_{nmax} = 0.332 (0.157 \times 10^6 mm^2) \sqrt{11.7 \frac{MN}{m^2}} \frac{1 kN}{1,000 N}$$
$$= 178 kN$$

V_n is less than this, so this equation isn't controlling.

Step 5d: Shear Capacity

Multiplying the nominal strength by the strength reduction factor we get:

$\phi V_n = 0.8(35.8k) \qquad\qquad \phi V_n = 0.8(160.8 kN)$
$ = 28.6k \qquad\qquad\qquad\;\; = 129 kN$

Step 6: Minimum Reinforcing Area

Next, we check the minimum shear reinforcing steel, as follows:

$A_{vmin} = 0.0007 bd_v$
$= 0.0007(7.625in)32in \qquad = 0.0007(194mm)810mm$
$= 0.171in^2 \qquad\qquad\qquad = 110mm^2$

Converting our shear steel to a per foot basis, we get:

$$A_{vft} = \frac{12}{8} A_v \qquad\qquad A_{vft} = \frac{305}{200} A_v$$

$$= \frac{12}{8}(0.11in^2) \qquad\qquad = \frac{305}{200}(71mm^2)$$

$$= 0.165in^2 \qquad\qquad\qquad = 108mm^2$$

This is close, but not quite there. Let's bump up to a #4 (M13) bar to be covered.

Step 7: Summarize Results

In summary, we have a 32 in (810 mm) deep lintel, with one #4 (M13) bar in every vertical cell (8 in, 200 mm, on center). It is grouted solid to provide adequate shear capacity and bond the reinforcing to the masonry.

5.7 WHERE WE GO FROM HERE

From here, there are two places we can go. First, we can calculate a higher masonry shear strength by looking at the moment–shear-depth ratio $M_u/(V_u d_v)$. TMS 402 provides for greater strength when this ratio is less than one. Second, if we prestress the masonry, we get additional shear strength, as the axial load increases.

NOTE

1. TMS, *Specification for Masonry Structures*, TMS 602–13 (Longmont, CO: The Masonry Society, 2013).

Masonry Compression

Chapter 6

Paul W. McMullin

6.1 Stability
6.2 Capacity
6.3 Demand versus Capacity
6.4 Deflection
6.5 Detailing Considerations
6.6 Slender Wall Example
6.7 Where We Go from Here

Columns, pilasters, and walls are the primary compression elements in masonry construction today. They keep everything up. Let's get into the basics of column design, then focus our efforts on slender masonry walls.

Key considerations for reinforced masonry compression design include

- Strength—Is the member large enough to carry axial and flexural loads?
- Stability—How prone is it to buckling?
- Layout—What is the column or pilaster spacing and floor heights?
- Detailing—How much reinforcing is required; how and where is it placed?

6.1 STABILITY

Buckling is a primary failure mode in column design of any material. Because columns and walls are braced at the floors, roof, and foundation, they are rather prone to buckling. If the column is slender, and loads are high enough, it will bow out in the middle between brace points (like the straw in Figure 6.1) and collapse. For a slender column, buckling

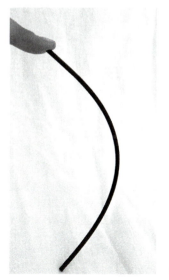

Figure 6.1 Buckled plastic straw

Masonry Compression

occurs at stresses less than the material strength. In compressing, solely focusing on strength can give us a false sense of safety.

We capture the propensity of a member to buckle by the slenderness ratio, given in the following equation:

$$\frac{h}{r} \tag{6.1}$$

where:

h = the member height, in (mm)
r = radius of gyration, in (mm)

The masonry code[1] compression strength equations are dependent on whether the slenderness ratio is above or below 99.

6.2 CAPACITY

The nominal axial capacity P_n of compression elements is given by the following two equations:

If $h/r \leq 99$,

$$P_n = 0.8 \left[0.8 f'_m (A_n - A_{st}) + f_y A_{st} \right] \left[1 - \left(\frac{h}{140r} \right)^2 \right] \tag{6.2}$$

If $h/r > 99$

$$P_n = 0.8 \left[0.8 f'_m (A_n - A_{st}) + f_y A_{st} \right] \left(\frac{70r}{h} \right)^2 \tag{6.3}$$

where:

f'_m = masonry compressive strength, k/in^2 (kN/m^2)
A_n = net cross-sectional area, in^2 (mm^2)
A_{st} = area of laterally tied reinforcing steel, in^2 (mm^2)
f_y = steel yield strength, k/in^2 (kN/m^2)
h = member height, in (mm)
r = least radius of gyration, in (mm)

Where columns experience compression and bending forces, we must consider these effects at the same time. This is because these forces cause stresses in the same direction and in the same reinforcing. We can use interaction diagrams to capture this effect, which plot axial strength versus flexural strength, shown in Figure 6.2. These diagrams capture the relative influence of moment on axial load capacity. We use them by

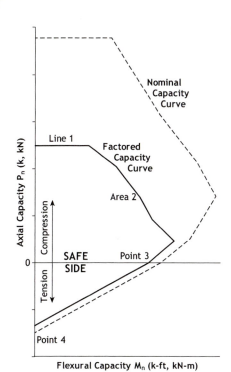

Figure 6.2 Example column interaction diagram

plotting the axial demand P_u and moment demand M_u. If the point is to the left of the curve, the column is strong enough.

Interaction diagrams are covered extensively in *Concrete Design*[2] in this series. It can be extended to masonry by using the two limiting equations above, and replacing f'_c with f'_m.

6.2.1 Slender Walls

A subset of masonry compression design is slender walls. They take advantage of reasonably low loads and reinforcing steel, typically in the middle of the wall, to resist axial and out-of-plane wind and seismic forces. We look at four parts: axial load, moment, axial strength, and moment strength.

Axial load in slender walls has two **components**: loads that are applied to the center of the wall P_{uw}—the wall weight—and loads applied eccentrically P_{uf}, floor and roofs. These are illustrated in Figure 6.3.

Masonry Compression

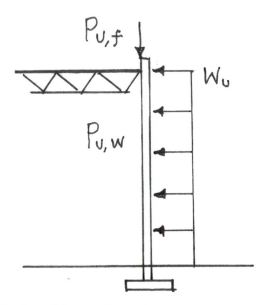

Figure 6.3 Slender wall forces and geometry

Moment has three components: out of plane forces, moments due to eccentric loads, and **P-delta** moments. The first two are shown in Figure 6.3, and given by the following equation.

$$M_{u,0} = \frac{w_u h^2}{8} + P_{uf} \frac{e_u}{2} \tag{6.4}$$

w_u = out-of-plane line load, k/ft (kN/m)
P_{uf} = tributary load from roof and floors, k (kN)
e_u = axial load eccentricity, in (mm)

P-delta moments account for increased bending from the deflected shape of the wall. As the wall displaces laterally, the axial load creates an additional moment, illustrated in Figure 6.4. The more the wall deflects, the more the moment increases. The wall either reaches equilibrium, where the strength is greater than the loads, or becomes unstable. We prefer the first option.

There are two options to calculate the bending moment in a slender wall. In the first, we calculate the deflection from bending forces and

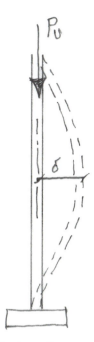

Figure 6.4 P-delta effect in a slender wall

initial eccentricity. Then calculate the increased moment, then new deflection and new moment. We do this until the deflections stops increasing.

The second method allows us to directly solve for the total, increased moment, as follows:

$$M_u = \Psi M_{u,0} \tag{6.5}$$

where:

$$\Psi = \frac{1}{\left(1 - \dfrac{P_u}{P_e}\right)} \tag{6.6}$$

P_u = factored axial load, k (kN)

$$P_e = \frac{\pi^2 E_m I_{eff}}{h^2} \tag{6.7}$$

Masonry Compression

E_m = masonry modulus of elasticity, k/in² (kN/m²)
I_{eff} = effective moment of inertia, in⁴ (mm⁴)
   = can be approximated as $0.5I_n$
I_n = net section moment of inertia
h = wall height, in (mm)

It can be a lot. Take your time, and work your way through each step.

We use equation (6.2) and (6.3) to calculate the axial capacity. Additionally, the code limits axial stress in slender walls to the following values.

For h/t ≤ 30 $\dfrac{P_u}{A_g} \leq 0.20 f'_m$ (6.8)

For h/t > 30 $\dfrac{P_u}{A_g} \leq 0.05 f'_m$

where:

P_u = factored axial load, k (kN)
A_g = gross cross-sectional area, in² (mm²)

To determine bending capacity, we use the information in Chapter 4. When calculating a, make sure it is within the face shell of the masonry unit. Otherwise, a portion of the stress will be carried by air, which doesn't work very well. Appendix 1 provides the face shell dimensions. Additionally, we limit the compression width b as follows:

- Center-to-center of the vertical bar spacing
- 6 times the nominal wall thickness
- 72 in (1,830 mm)

BOX 6.1 INITIAL WALL SIZING

For slender walls up to 26 ft (7.9 m) tall, 8 in (203 mm) block tends to work reasonably well, as indicated in Table 6.1. This table also provides an estimate of reinforcing steel. Slender walls are most influenced by lateral loads (wind and seismic), and less by axial load. Keep in mind, that you may need a thicker wall, or pilasters, at large openings or at heavy point loads.

Table 6.1 Initial slender wall sizes

	Imperial			Metric		
Bearing	Wall	Reinforcing		Wall	Reinforcing	
Height	Thick	Size	Spacing	Thick	Size	Spacing
(ft)	(in)		(in)	(mm)		(mm)
10	8	#4	48	200	M13	1,220
12	8	#4	48	200	M13	1,220
14	8	#5	48	200	M16	1,220
16	8	#5	32	200	M16	820
18	8	#5	32	200	M16	820
20	8	#5	24	200	M16	610
22	8	#6	24	200	M19	610
24	8	#6	24	200	M19	610
26	8	#7	24	200	M22	610
30	10	#7	24	250	M22	610
34	10	#7	16	250	M22	410
40	12	#7	16	305	M22	410

Notes: 1) This table is for initial sizing only. Final size must be determined by a licensed structural engineer.
2) Table assumes one single bar centered in the wall. Higher capacities are available with two bars, closer to the face.
3) Based on 20 lb/ft² (0.96 kN/m²) dead, 30 lb/ft² (1.44 kN/m2) snow, and 45 lb/ft² (2.16 kN/m²) lateral wind loads, and 20 ft (6.1 m) tributary length.

6.3 DEMAND VERSUS CAPACITY

Like previous chapters, when $\phi P_n \geq P_u$ our column has sufficient capacity. For walls we add $\phi M_n \geq M_u$ to the check. ϕ is 0.90 for axial load and bending moment. If these relationships are not met, we can add reinforcing, make the member larger, or find a way to reduce loads.

6.4 DEFLECTION

Axial deflection is not a concern in masonry columns. However, in slender walls, deflection can be a problem. We limit the factored, mid-height deflection Δ_u of walls to $0.01h$.

Masonry Compression

We calculate slender wall deflection as follows:

When $M_u \leq M_{cr}$
$$\delta_u = \frac{5M_u h^2}{48 E_m I_n} \tag{6.9}$$

When $M_u > M_{cr}$
$$\delta_u = \frac{5M_{cr} h^2}{48 E_m I_n} + \frac{5(M_u - M_{cr}) h^2}{48 E_m I_{cr}} \tag{6.10}$$

where:

M_u = moment on the wall, k-ft (kN-m), from equation (6.5)
h = wall height, ft (m)
E_m = masonry modulus of elasticity, k/in² (kN/m²)
I_n = net moment of inertia, in⁴ (mm⁴)
I_{cr} = cracked moment of inertia, in⁴ (mm⁴)
   = $0.4 I_n$
M_{cr} = cracking moment, k-ft (kN-m), from equation $M_{cr} = f'_r S_x$

Once we get the mid-height wall deflection we compare it to the allowable Δ_a.

6.5 DETAILING CONSIDERATIONS

Strength and deformability of masonry columns and walls depends on how well they are detailed and constructed. In a lower load wall, like the one in the example at the end of the chapter, vertical bars are spaced at regular intervals, with additional bars near openings, shown in Figure 6.5. Where additional capacity is needed in a wall, we add **hoops** to confine the masonry, shown in Figure 6.6.

Figure 6.5 Masonry wall construction showing vertical wall and column bars

Figure 6.6 Confinement ties in a flush, masonry column

TMS 402 provides requirements for vertical bars and ties in columns. These are provided in Figure 6.7 and Figure 6.8, respectively. They ensure adequate cover, grout engagement, strength, and ductility.

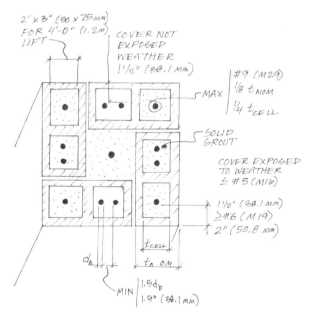

Figure 6.7 Column lengthwise bar requirements

Masonry Compression

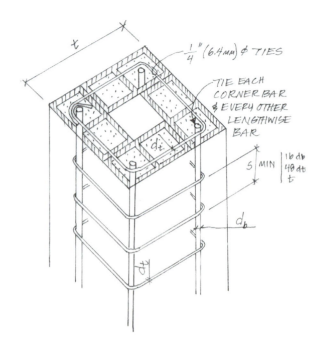

Figure 6.8 Column **tie** requirements

> **BOX 6.2 DESIGN STEPS**
>
> The following steps will guide you through the design of a slender wall.
>
> 1. Draw the structural layout, include height and tributary width
> 2. Determine loads—Unit, load combinations generating a line load, wall axial force, and bending moment
> 3. Material parameters—Choose masonry compressive strength and reinforcing strength
> 4. Estimate initial masonry size and reinforcing
> 5. Determine the axial and bending strengths
> 6. Check minimum reinforcing steel
> 7. Check deflection
> 8. Summarize the results

6.6 SLENDER WALL EXAMPLE

Step 1: Structural Layout

Continuing with the small maintenance shop in Chapter 4, we now design the wall for axial load and bending. We will use the slender wall provisions of TMS, discussed in Section 6.2.1.

Based on Figure 4.8, we will take the wall and parapet wall heights as

h_w = 15ft $\qquad\qquad h_w = 4.57m$

h_p = 2ft $\qquad\qquad h_p = 0.61m$

and the tributary roof length as

l_t = 19.33ft $\qquad\qquad l_t = 5.89m$

With this information, we draw the **free body diagram** of the wall, shown in Figure 6.9. This gives us a simple way to keep track of all the geometric information and loading we will need.

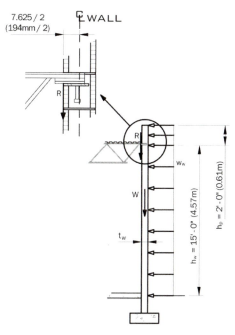

Figure 6.9 Free body diagram for slender wall example

Masonry Compression

Step 2: Determine the Loads

Building off the information in the example for Chapter 4, our roof, dead, and snow loads are as follows:

$$D = 20\frac{lb}{ft^2} \qquad\qquad D = 0.96\frac{kN}{m^2}$$

$$S = 50\frac{lb}{ft^2} \qquad\qquad S = 2.39\frac{kN}{m^2}$$

The unit wall weight, for an 8 in (200 mm) thick wall, from Table 2.4. Note, we use the solid grouted value, in case we need to space the reinforcing closer.

$$D_w = 78\frac{lb}{ft^2} \qquad\qquad D_w = 3.73\frac{kN}{m^2}$$

For wall weight, we take the parapet height plus half the wall height from roof to floor, as the highest moment is at the mid-height of the wall. We now calculate the loads to the wall, assuming a spacing *s* of 1 ft (1 m) width. (Note this will cause most of the numbers below to not convert directly between Imperial and Metric units.) We will keep the roof load separate from the wall load, as it is eccentric to the wall, and creates an additional bending moment.

We will use the seismic dominant load combination for this example. However, it would also be wise to check the snow load dominant combination. Perhaps you can do this on your own.

Factored	*Unfactored*
1.2D+1.0E+0.2S	D+0.525E+0.75S

Roof loads are as follows:

$$R_D = Dl_t s$$
$$= 20\frac{lb}{ft^2}(19.33ft)1ft$$
$$= 387\,lb$$

$$= 0.96\frac{kN}{m^2}(5.89m)1m$$
$$= 5.65\,kN$$

$$R_S = Sl_t s$$
$$R_s = 50\frac{lb}{ft^2}(19.33ft)1ft$$
$$= 967\,lb$$

$$= 2.39\frac{kN}{m^2}(5.89m)1m$$
$$= 14.1\,kN$$

The wall **dead load** at the mid-height is:

$$W_D = D_w \left(h_p + \frac{h_w}{2} \right) s$$

$$= 0.078 \frac{k}{ft^2} \left(2ft + \frac{15ft}{2} \right) 1ft \qquad = 3.73 \frac{kN}{m^2} \left(0.61m + \frac{4.57m}{2} \right) 1m$$

$$= 0.74 k \qquad\qquad\qquad\qquad\qquad = 10.8 kN$$

Now, we calculate the out of plane forces due to seismic forces. The out-of-plane wall force is given in ASCE 7 as:

$F_p = 0.4 I_E S_{DS} w_w$

$I_E = 1.0$, based on occupancy

$S_{DS} = 1.25$, based on site seismicity

$$w_w = D_w s$$

$$= 78 \frac{lb}{ft^2} (1ft) \qquad\qquad = 3.73 \frac{kN}{m^2} (1m)$$

$$= 78 \frac{lb}{ft} \qquad\qquad\qquad = 3.73 \frac{kN}{m^2}$$

$$F_P = 0.4(1.0)1.25 \left(78 \frac{lb}{ft} \right) \qquad F_P = 0.4(1.0)1.25 \left(3.73 \frac{kN}{m} \right)$$

$$= 39 \frac{lb}{ft} \qquad\qquad\qquad\qquad = 1.87 \frac{kN}{m}$$

The seismic term E has a vertical and horizontal component in it. The horizontal is given as F_p. The vertical is defined as $0.2 S_{DS} D$, which gives $0.2(1.25) = 0.25$. We add these to the dead term in the load combinations above, which now become:

Factored	Unfactored
1.45D+1.0E+0.2S	1.25D+0.525E+0.75S

This is a good time to remember that load combinations often don't add up to a single number. In this case, we have forces in two perpendicular directions. We therefore keep the axial and **seismic loads** separated. The unfactored and factored axial loads are as follows:

Masonry Compression

Unfactored

$R = R_D + R_S$
  $= 1.25(0.387k) + 0.75(0.967k)$ $= 1.25(5.65kN) + 0.75(14.1kN)$
  $= 1.21k$ $= 17.6 kN$

Factored

$R_u = 1.45 R_D + 0.2 R_S$
  $= 1.45(0.387k) + 0.2(0.967k)$ $= 1.45(5.65kN) + 0.2(14.1kN)$
  $= 0.75k$ $= 11.0 kN$

$W_{Du} = 1.45 W_D$
  $= 1.45(0.74k)$ $= 1.45(10.8kN)$
  $= 1.07k$ $= 15.7 kN$

We use factored combinations for strength calculations and unfactored for deflection.

We still need to calculate bending moment in the wall. However, it would help to have our material parameters and initial member size.

Step 3: Masonry Parameters

Building off the example in Chapter 4, we have:

$$f'_m = 1{,}700 \frac{lb}{in^2} \qquad\qquad f'_m = 11.7 \frac{MN}{m^2}$$

From this, we can calculate the modulus of elasticity:

$E_m = 900 f'_m$
  $= 900\left(1.7 \frac{k}{in^2}\right)$ $= 900\left(11.7 \frac{MN}{m^2}\right)$
  $= 1{,}530 \frac{k}{in^2}$ $= 10{,}530 \frac{MN}{m^2}$

We are using running bond. Thus, from Table 2.6 the modulus of rupture is:

$$f_r = 84 \frac{lb}{in^2} \qquad\qquad f_r = 579 \frac{kN}{m^2}$$

The reinforcing steel yield strength is:

$$f_y = 60\frac{k}{in^2} \qquad f_y = 420\frac{MN}{m^2}$$

Step 4: Initial Size

Our wall thickness of 8 in (200 mm) is locked in from our lintel design. This gives:

$$t_w = 7.625 in \qquad t_w = 194 mm$$
$$b = 12 in \qquad b = 1{,}000 mm$$

We can get a sense for the required reinforcing from Table 6.1. To be on the safe side, we will try a #6 (M19) bar at 24 in (610 mm) on center, giving us an area of:

$$A_{st} = 0.44 in^2 \qquad A_{st} = 284 mm^2$$

Because the bar isn't spaced at 12 in (305 mm), we need to convert this area to a unit basis.

$$A_{st} = 0.44 in^2 \left(\frac{12 in}{24 in}\right) \qquad A_{st} = 284 mm^2 \left(\frac{1{,}000 mm}{610 mm}\right)$$
$$= 0.22 in^2 \qquad\qquad = 466 mm^2$$

From Table 2.9, we see that our chosen bar size is smaller than the limit. Cool!

Step 5: Member Strength

Atypically, we start with finding the bending moment in the wall, at the mid-height, using the following equation.

$$M_u = \frac{F_P h_w^2}{8} + R_u \frac{e_u}{2} + P_u \delta_u$$

We have most of these terms. e_u is the eccentricity of the roof loads to the wall. For simplicity, we will assume the roof **reaction** is applied at the face of the wall (see Figure 6.9).

$$e_u = \frac{7.625 in}{2} \qquad e_u = \frac{194 mm}{2}$$
$$= 3.81 in \qquad\qquad = 97 mm$$

P_u is the total axial load, given as follows:

$$P_u = R_u + W_{Du}$$
$$= 0.75k + 1.07k \qquad\qquad = 11.0kN + 15.7kN$$
$$= 1.82k \qquad\qquad\qquad\quad\; = 26.7kN$$

Now to find the moment in the wall, we will start with the first two terms of the M_u equation above. Then we will calculate δ_u, and begin our iterations.

$$M_u = \frac{0.039\frac{k}{ft}(15ft)^2}{8} + 0.75k\frac{3.81in}{2}\left(\frac{1ft}{12in}\right) + 0$$
$$= 1.22k-ft$$

$$M_u = \frac{1.87\frac{kN}{m}(4.57m)^2}{8} + 11.0kN(0.097m) + 0$$
$$= 5.42kN-m$$

To calculate deflections, we will need to first find the cracking moment, given as:

$$M_{cr} = S_{avg}f_r$$

From Table A1.1 in Appendix 1, we get the section modulus perpendicular to the **bed joints** on a unit width basis. For running bond:

$$S_{avg} = 96.9in^3 \qquad\qquad S_{avg} = 5{,}210 \times 10^3 mm^3$$
$$M_{cr} = 96.9in^3\left(0.084\frac{k}{in^2}\right)\frac{1ft}{12in} \quad M_{cr} = 5{,}210 \times 10^3 mm^3\left(\frac{1m}{1{,}000mm}\right)^3\left(579\frac{kN}{m^2}\right)$$
$$= 0.68k-ft \qquad\qquad\qquad\quad = 3.0kN-m$$

Because the moment demand is greater than the modulus of rupture, we use equation (6.10).

$$\delta_u = \frac{5M_{cr}h_w^2}{48E_m I_n} + \frac{5(M_u - M_{cr})h_w^2}{48E_m I_{cr}}$$

Also from Table A1.1, the net and cracked moment of inertia are:

$$I_n = 369in^4 \qquad\qquad I_n = 504 \times 10^6 mm^4$$
$$I_{cr} = 0.4I_n$$

$$= 0.4(369 in^4)$$
$$= 147.6 in^4$$

$$= 0.4(504 \times 10^6 mm^4)$$
$$= 202 \times 10^6 mm^4$$

With these, we can calculate deflection for the first time.

$$\delta_u = \left[\frac{5(0.68k-ft)(15ft)^2}{48\left(1,530\frac{k}{in^2}\right)369 in^4} + \frac{5(1.22-0.68)(15ft)^2}{48\left(1,530\frac{k}{in^2}\right)148 in^4}\right]\left(\frac{12 in}{1 ft}\right)^3$$
$$= 0.145 in$$

$$\delta_u = \left[\frac{5(3.0 kN-m)(4.57m)^2}{48\left(10.5\frac{kN}{mm^2}\right)504 \times 10^6 mm^4} + \frac{5(5.42-3.0)(4.57m)^2}{48\left(10.5\frac{kN}{mm^2}\right)202 \times 10^6 mm^4}\right]\left(\frac{1,000 mm}{1 m}\right)^3$$
$$= 3.72 mm$$

With this, we can start our iteration, by calculating the moment again, this time using all three terms.

$$M_u = \frac{0.039\frac{k}{ft}(15ft)^2}{8} + \left[0.75k\frac{3.81 in}{2} + 1.82k(0.145 in)\right]\left(\frac{1 ft}{12 in}\right)$$
$$= 1.24 k-ft$$

$$M_u = \frac{1.87\frac{kN}{m}(4.57m)^2}{8} + 11.0 kN\left(\frac{0.097m}{2}\right) + 26.7 kN(0.0037m)$$
$$= 5.51 kN-m$$

And deflection again:

$\delta_u = 0.149 in$ $\qquad$ $\delta_u = 3.79 mm$

Then moment again:

$M_u = 1.24 k\text{--}ft$ $\qquad$ $M_u = 5,42 kN-m$

And deflection again:

$\delta_u = 0.149 in$ $\qquad$ $\delta_u = 3.79 mm$

At this point, we see the deflection and moment aren't changing. This indicates that we have reached equilibrium and that the wall is stable.

Masonry Compression

With this, we can check strength. Let's begin with axial compression strength. First, we determine whether we should use equation (6.2) or (6.3). We do this by calculating the slenderness ratio. We will need the radius of gyration:

$r = 2.53 in$

$\dfrac{h_w}{r} = \dfrac{15 ft}{2.53 in}\left(\dfrac{12 in}{ft}\right)$

$= 71$

$r = 64.3 mm$

$\dfrac{h_w}{r} = \dfrac{4{,}570 mm}{64.3 mm}$

$= 71$

Axial strength is thus given by equation (6.2), as:

$$P_n = 0.8\left(0.8 f'_m (A_n - A_{st}) + f_y A_{st}\right)\left[1 - \left(\dfrac{h_w}{140 r}\right)^2\right]$$

We will need both net and gross area. From Table A1.1:

$A_n = 51.3 in^2$
$A_g = b t_w$
$\quad = 12 in (7.625 in)$
$\quad = 91.5 in^2$

$A_n = 109 \times 10^3 mm^2$

$\quad = 1{,}000 mm (194 mm)$
$\quad = 194{,}000 mm^2$

Axial strength follows as:

$$P_n = 0.8\left(0.8\left(1.7\dfrac{k}{in^2}\right)(51.3 in^2 - 0.22 in^2) + 60\dfrac{k}{in^2}(0.22 in^2)\right)\left[1 - \left(\dfrac{15 ft (12 in / 1 ft)}{140 (2.53 in)}\right)^2\right]$$

$= 49 k$

$$P_n = 0.8\left(0.8\left(0.0117\dfrac{k}{mm^2}\right)(109 \times 10^3 - 466) mm^2 + 0.42\dfrac{kN}{mm^2}(466 mm^2)\right)\left[1 - \left(\dfrac{4{,}570 mm}{140 (64.3 mm)}\right)^2\right]$$

$= 720 kN$

Applying the strength reduction factor of:

$\phi = 0.9$

We get:

$\phi P_n = 44.1 k$ $\phi P_n = 648 kN$

This is much larger than the axial demand P_u. Cool!

We now check the limiting axial stress. Because $h_w/r > 30$, the axial stress must be lower than $0.05 f'_m$.

$0.05 f_m = 0.05 (1,700 \frac{lb}{in^2}) = 85 \frac{lb}{in^2}$

$0.05 f_m = 0.05 (11,700 \frac{kN}{m^2}) = 585 \frac{kN}{m^2}$

Our axial stress is low enough to use this method, so we are good to move forward and check bending strength, like in Chapter 4.

$\frac{P_u}{A_g} = \frac{1,820 lb}{91.5 in^2} = 20 \frac{lb}{in^2}$ $\frac{P_u}{A_g} = \frac{26.7 kN}{0.194 m^2} = 138 \frac{kN}{m^2}$

Because the bar is centered in the wall, the distance from the reinforcing to top of lintel is half the wall thickness.

$d = \frac{t_w}{2}$
$= \frac{7.625 in}{2} = 3.81 in$ $= \frac{194 mm}{2} = 97 mm$

We next find the tension force in the reinforcing, and compression block depth.

$T = A_{st} f_y$
$= 0.22 in^2 \left(60 \frac{k}{in^2} \right)$
$= 13.2 k$

$= 466 mm^2 \left(0.42 \frac{kN}{mm^2} \right)$
$= 196 kN$

$a = \frac{T}{0.8 f'_m b}$

$= \frac{13.2 k}{0.8 \left(1.7 \frac{k}{in^2} \right) 12 in}$

$= 0.81 in$

$= \frac{196 kN}{0.8 \left(0.0117 \frac{kN}{mm^2} \right) 1,000 mm}$

$= 20.9 mm$

Masonry Compression

The nominal bending capacity is follows as:

$$M_n = T\left(d - \frac{a}{2}\right)$$

$$= 13.2k\left(3.81in - \frac{0.81in}{2}\right)\frac{1ft}{12in} \qquad = 196kN\left(97mm - \frac{20.9mm}{2}\right)\frac{1m}{1,000mm}$$

$$= 3.75k-ft \qquad\qquad\qquad\qquad\qquad = 17.0kN-m$$

Multiplying by a strength reduction factor ϕ of 0.9, we get:

$$\phi M_n = 3.38k-ft \qquad\qquad\qquad \phi M_n = 15.3kN-m$$

This is greater than the demand M_u, which tells us our design works! Because this is about double the demand, we could try a smaller bar size, or larger spacing.

Step 6: Minimum Reinforcing

Checking minimum reinforcing, we see M_n is larger than 1.3 M_{cr}. We have enough reinforcing steel to meet the code minimum.

Step 7: Check Deflection

The allowable deflection is:

$$\delta_{all} = 0.01 h_w$$

$$= 0.01(15ft)\frac{12in}{1ft} \qquad\qquad = 0.01(4,570mm)$$

$$= 1.8in \qquad\qquad\qquad\qquad = 46mm$$

This is much larger than the factored deflection δ_u we calculated above in our iterations, which is higher than the unfactored deflection will be.

Step 8: Summary

In summary, we have an 8 in (200 mm) CMU wall with a #6 vertical bar spaced at 24 in (610 mm) on center. Perhaps you can recalculate the example with a #5 (M16) at 32 in (810 mm) on center.

6.7 WHERE WE GO FROM HERE

We have looked at the basic requirements for axial compression and slender wall designs. From here, when we have combined axial and bending loads in columns, we can use interaction diagrams. This brings us into a set of more rigorous design tools.

NOTES

1. TMS, *Building Code Requirements for Masonry Structures*, TMS 402–13 (Longmont, CO: The Masonry Society, 2013).

2. P.W. McMullin, J.S. Price, E. Persellin *Concrete Design*, Architects Guidebook to Structures (New York: Routledge, 2016).

Timber Lateral Design

Chapter 7

Paul W. McMullin

7.1 Introduction
7.2 Lateral Load Paths
7.3 Diaphragms
7.4 Shear Walls
7.5 Seismic Design Considerations
7.6 Where We Go from Here

7.1 INTRODUCTION

Lateral loads on structures are commonly caused by wind, earthquakes, and soil pressure, and less commonly from human activity, waves, or blasts. These loads are difficult to quantify with any degree of precision. However, following reasonable member and system proportioning requirements, coupled with prudent detailing, we can build reliable masonry structures that effectively resist lateral loads.

What makes a structure perform well in a windstorm is vastly different than an earthquake. A heavy, squat structure, such as the Parthenon in Greece can easily withstand wind—even without a roof. Its mass anchors it to the ground. On the other extreme, a tent structure could blow away in a moderate storm. Conversely, the mass of the Parthenon makes is extremely susceptible to earthquakes (remember earthquake force is a function of weight), while the tent in a seismic event will hardly notice what is going on. Masonry tends to be more sensitive to seismic forces, than wind.

Looking at this closer, wind forces are dependent on three main variables:

1) Proximity to open spaces such as water or mud flats
2) Site exposure
3) Building shape and height.

In contrast, earthquake forces are dependent on very different variables:

4) Nature of the seismic event
5) Building weight
6) **Rigidity** of the **structural system**.

Because we operate in a world with gravity forces, we inherently understand the **gravity load paths** of the simple building shown in Figure 7.1a. Downward loads enter the roof and floors and make their way to the walls, columns, and eventually footings. Lateral loads can take more time to grasp. But we can think of them as turning everything 90 degrees; the structure acting as a cantilevered beam off the ground, illustrated in Figure 7.1b.

The magnitude and distribution of lateral loads drives the layout of shear walls. These walls resist lateral forces, acting like cantilevered beams poking out of the ground.

Timber Lateral Design

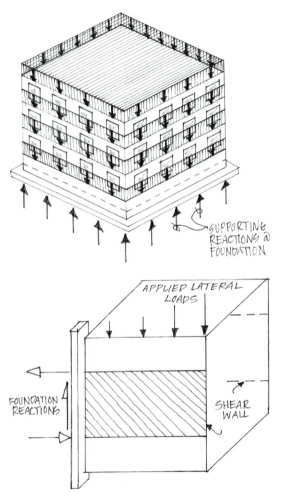

Figure 7.1 (a) Gravity load path, (b) lateral load path turned 90 degrees

We design lateral wind resisting members to not damage the system. Conversely, because strong seismic loads occur much less frequently, we design their lateral systems to yield the steel reinforcing. This absorbs significant amounts of energy, as illustrated in Figure 7.2, resulting in smaller member sizes. However, it leaves the structure damaged.

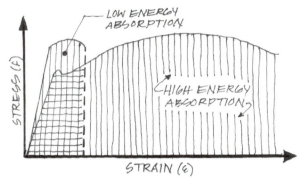

Figure 7.2 Comparative energy absorption for high and low deformation behavior

For design of seismic load resisting systems, we follow rigorous member proportioning and detailing requirements to ensure yielding occurs in the right places. This chapter focuses on design and detailing requirements from a conceptual point of view, and what lateral load resisting systems, elements, and **connections** should look like.

7.2 LATERAL LOAD PATHS

Following the path lateral loads travel through a structure is key to logical structural configuration and detailing. If the load path is not continuous from the roof to the ground, failure can occur. Additionally, no amount of structural engineering can compensate for an unnecessarily complex load path.

When configuring the structure, visualize how lateral forces—and gravity forces—travel from element to element, and eventually to the ground. A well-planned load path will save weeks of design effort, substantially reduce construction cost, and minimize structural risk. Software can't do this, but careful thought will.

Looking at lateral load paths further, Figure 7.3 shows how they enter a structure and find their way to the ground. Starting at point 1, wind induces pressure, or seismic accelerations cause inertial forces, perpendicular to the face of the building. Spanning vertically (point 2), the wall delivers a line or point load to a connection at the roof or floor level. The roof or floor picks up additional inertial seismic load. The roof (number 3) must resist lateral forces through diaphragm

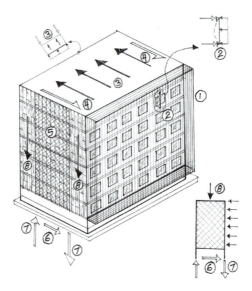

Figure 7.3 Detailed lateral load path in structure

action—essentially a deep beam. The ends of the diaphragm (point 4) then deliver load into connections to the shear walls. This occurs at each level (point 5). The lateral force works its way to the footing (point 6), which transfers the force to the soil through friction and passive pressure. Because the lateral forces are applied at a distance above the ground, they impart an overturning moment to the system. This causes tension and compression in the ends of shear walls and outside frame columns (point 7). The weight of the structure (point 8) helps resist this overturning moment, keeping it from tipping over.

To review, lateral loads are applied perpendicular to walls or cladding. Bracing these are the roof and floor diaphragms, which transfer their loads to the walls parallel to the load. Walls are supported by the ground. The weight of the structure (and sometimes deep foundations) keeps the system from tipping over.

Connections are critical to complete load paths. We need to ensure the lateral loads flow from perpendicular wall and floor, into diaphragms, into walls parallel to the load, and down to the foundation. Each time the load enters a new element, there must be a connection.

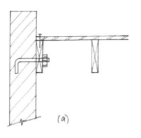

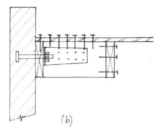

Figure 7.4 (a) Ineffective, and (b) effective concrete wall to wood roof connection

Connections from light to heavier materials warrant special consideration, particularly roof to wall interfaces. Figure 7.4a shows a common, seismically deficient, connection between a masonry wall and wood roof. As the wall moves away from the sheathing, the 2x is placed in cross-grain bending, resulting in failure. Without much additional effort, we can connect the wall to a metal strap and blocking like Figure 7.4b and get a connection that will keep the wall from tipping over.

7.3 DIAPHRAGMS

Lateral systems include horizontal and vertical elements. Horizontal systems consist of diaphragms and **drag struts** (**collectors**). Vertical elements consist of shear walls and frames. Horizontal systems transfer forces through connections to vertical elements, which carry the loads into the foundation.

Diaphragms consist of structural panels, and straight or diagonal sheathing boards. In other structures they are made of concrete slabs, bare metal deck, and diagonal bracing. Timber diaphragms have comparatively low capacity. However, for smaller structures, or those with light walls, timber diaphragms perform well. Diaphragms make possible large open spaces, without internal walls—so long as there is adequate vertical support.

7.3.1 Forces

We can visualize diaphragms as deep beams that resist lateral loads, illustrated in Figure 7.5a. They experience maximum bending forces near their middle, and maximum shear at their supports (where they connect to walls or frames), as seen in Figure 7.5b.

We resolve the mid-span moments into a tension–compression couple, requiring **boundary elements** around their edges. These are typically bond beams and can often resist these forces with a single bar, since the distance between them is large.

Shear forces are distributed throughout the length of the diaphragm in the direction of lateral force. Because many shear walls and frames

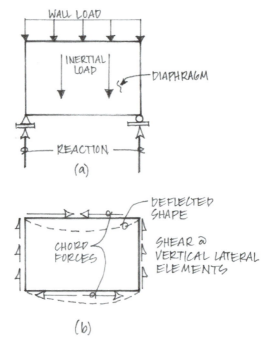

Figure 7.5 (a) Diaphragm forces and reactions, and (b) internal forces

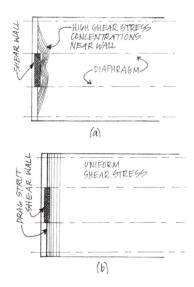

Figure 7.6 Diaphragm stress distribution (a) without, and (b) with drag struts

do not go the length of the building, the transfer of shear forces between the diaphragm and vertical elements causes high stress concentrations at the ends of the wall or frame, illustrated in Figure 7.6a. By adding drag struts (also called collectors), we gather the shear stresses into this stronger element, which can then deliver the force to the wall or frame. This reduces the stress concentration (Figure 7.6b) and ensures the diaphragm retains its integrity. Drag struts frequently consist of beams, joists, and metal straps in timber structures. Note that a structural element that acts as drag strut will act as a diaphragm chord when the forces are turned and analyzed 90 degrees.

Timber Design in this series, Chapter 8, thoroughly discusses timber diaphragms. *Concrete Design* and *Steel Design* have similar sections discussing concrete and metal deck diaphragms, respectively.

7.4 SHEAR WALLS

Reinforced masonry shear walls are the most common vertically oriented lateral systems in masonry structures today. In older structures, we frequently find unreinforced masonry walls, where the brick is connected solely with mortar.

7.4.1 Forces

Shear walls resist lateral loads through horizontal shear along the length of the wall and tension–compression couples at the end, shown in Figure 7.7. They are efficient and stiff, and particularly suited to buildings with partitions and perimeter walls without an overabundance of windows. Short, tall shear walls concentrate forces at their base, requiring large foundations, while long, short walls reduce footing loads.

7.4.2 Analysis

To analyze a shear wall, we need to know how the lateral forces are distributed between wall segments in the same wall. While we could base it on wall length, we typically use stiffness. Stiffness is a measure of force required to deflect a member a unit distance—giving units of k/in (kN/m). The stiffness k of a cantilevered and fixed (**pier**) wall segment is given as follows:

$$k_{CANT} = \frac{1}{\dfrac{h^3}{3E_m I_g}} + \frac{1}{\dfrac{h}{A_v G_m}} \qquad k_{PIER} = \frac{1}{\dfrac{h^3}{12E_m I_g}} + \frac{1}{\dfrac{h}{A_v G_m}} \tag{7.1}$$

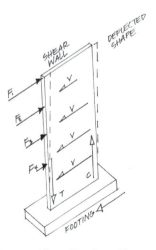

Figure 7.7 Shear wall external lateral loads and internal forces

where:

    h = **effective height** of wall, in (mm)
    E_m = modulus of elasticity of masonry, k/in² (MN/mm²)
    I_{cr} = moment of inertia of the wall, in⁴ (mm⁴), taken as $0.4I_g$
    A_v = wall area in direction of force, in² (mm²)
    G_m = shear modulus of elasticity of masonry, k/in² (MN/mm²), taken as $0.4E_m$

With the stiffness of each wall, we find the force to a single segment V_x by dividing the stiffness of the segment by the sum of all segment stiffnesses, then multiplying this by the force applied to the entire wall. In equation form, this is:

$$V_x = V \frac{k_x}{\sum k_{xi}} \tag{7.2}$$

where:

    V = shear force applied to the wall, k (kN)
    k_x = stiffness of the wall in question, k/in (kN/mm)
    k_{xi} = stiffness of each segment of wall, k/in (kN/mm)

Once we determine the forces in each segment, we can analyze them as stand-alone walls, as discussed in the next paragraphs.

In a shear wall the shear force is constant from top to bottom. The moment is a maximum at the bottom and zero at the top. Figure 7.8 illustrates the applied forces, and shear and moment diagrams.

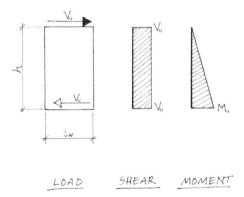

Figure 7.8 Shear wall shear and moment diagrams and design forces

The steps to analyze a segment of shear wall are as follows:

- Draw the wall with the applied shear force, reactions, and dimensions h and l_w
- Calculate the moment at the base of the wall as:

$$M_u = V_u h \tag{7.3}$$

where:

V_u = shear force at the top of the wall, lb (kN)
h = wall height, ft (m)

7.4.3 Capacity

Knowing the shear and moment in the wall, we next find the capacity of the wall. We use Chapter 4 to determine bending strength, and Chapter 5 to calculate shear capacity.

7.5 SEISMIC DESIGN CONSIDERATIONS

Seismic design centers on yielding specific members in the structure to absorb energy. The more it yields, the more energy is absorbed, as indicated in Figure 7.9. This reduces member sizes and creates more economical structures. In shear walls, we yield the tension reinforcing near the ends of the walls.

Building codes limit different seismic systems and their maximum heights to ensure the structures perform well during an earthquake. Table 7.1 presents different timber lateral systems and their permissible heights. It also includes the Response Modification Factor R, which

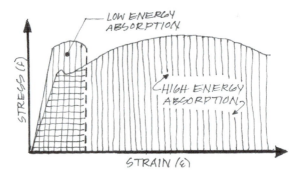

Figure 7.9 Low and high energy absorption in different materials

Table 7.1 Seismic lateral system R factors and maximum heights

Seismic Force Resisting System	Response Coefficient R	Permitted Height Seismic Category				
		B	C	D	E	F
Masonry						
Special Reinforced Shear Walls	5	NL	NL	160	160	100
Intermediate Reinforced Shear Walls	3 1/2	NL	NL	NP	NP	NP
Ordinary Reinforced Shear Walls	2	NL	160	NP	NP	NP
Ordinary Plain Shear Walls	1 1/2	NL	NP	NP	NP	NP
Timber						
Light Frame Walls Structural Panel Sheathed Walls	6 1/2	NL	NL	65	65	65
Concrete						
Special Moment Frames	8	NL	NL	NL	NL	NL
Special Reinforced Shear Walls	5	NL	NL	160	160	100
Steel						
Special Moment Frames	8	NL	NL	NL	NL	NL
Special Concentrially **Braced Frames**	6	NL	NL	160	160	100

Source: ASCE 7–10
NL= No Limit
NP= Not Permitted

reduces the seismic design force, as a function of energy absorption. A higher R indicates a better performing seismic system.

7.5.1 Response Modification

Because we design seismic systems to yield, the building code permits us to reduce the design seismic force. We do this by dividing it by the Response Modification Factor R, which is a function of energy absorption. A higher R indicates a better performing seismic system. Table 7.1 provides these for various lateral force resisting systems.

Codes also limit the height of most lateral systems in high seismic regions. These limits are based on seismicity and manifest themselves as **seismic design categories** B through F, listed in Table 7.1. Category D is the most common in regions of high seismicity. Categories E and F apply to very high seismicity, and buildings with higher societal importance.

7.5.2 Drift

For seismic forces, the code limits how much relative movement is permissible between floors—known as **drift** Δ. Think of it like a stack of dinner plates sliding off each other. Limiting drift helps reduce damage to cladding, partitions, mechanical ducts, and plumbing. Drift is determined from stiffness and load, and compared to the limits shown in Table 7.2—which are a function of story height.

7.5.3 Configuration Requirements

Building configurations that have horizontal jogs, vertical steps, large diaphragm openings, or large stiffness changes perform less effectively than their counterparts. This is because force concentrates in sharp changes of geometry, and the load path through these is inefficient. Examples of horizontal and vertical irregularities are shown in Figure 7.10 and Figure 7.11, respectively, along with potential options to avoid them.

Expanding further, horizontal structural irregularities include:

- Torsion, which occurs where there is a substantial difference in lateral system stiffness, such as a building with shear walls

Table 7.2 **Drift limits for multi-story structures**

Drift Limits			
	Risk Category		
Structural System	*I or II*	*III*	*IV*
Structures 4 stories or less, non masonry, with interior walls & ceilings designed to accommodate drift	$0.025h_{sx}$	$0.020h_{sx}$	$0.015h_{sx}$
Masonry cantilever shear walls structures	$0.010h_{sx}$	$0.010h_{sx}$	$0.010h_{sx}$
Other masonry shear wall structures	$0.007h_{sx}$	$0.007h_{sx}$	$0.007h_{sx}$
All other structures	$0.020h_{sx}$	$0.015h_{sx}$	$0.010h_{sx}$

Source: ASCE 7–10

h_{sx} = Story height under level being considered, don't forget to convert to inches or mm

Table 7.2 Continued

		Allowable Drift Values for Various Criteria & Lengths						
		Story Height (ft)						
Limit Criteria		8	10	12	14	15	16	20
		Allowable Drift Δ_a (in)						
0.007	h_{sx}	0.67	0.84	1.01	1.18	1.26	1.34	1.68
0.010	h_{sx}	0.96	1.20	1.44	1.68	1.80	1.92	2.40
0.015	h_{sx}	1.44	1.80	2.16	2.52	2.70	2.88	3.60
0.020	h_{sx}	1.92	2.40	2.88	3.36	3.60	3.84	4.80
0.025	h_{sx}	2.40	3.00	3.60	4.20	4.50	4.80	6.00
		Story Height (m)						
Limit Criteria		2.5	3.0	3.5	4.0	4.5	5.5	6.0
		Allowable Drift Δ_a (mm)						
0.007	h_{sx}	17.5	21.0	24.5	28.0	31.5	38.5	42.0
0.010	h_{sx}	25.0	30.0	35.0	40.0	45.0	55.0	60.0
0.015	h_{sx}	37.5	45.0	52.5	60.0	67.5	82.5	90.0
0.020	h_{sx}	50.0	60.0	70.0	80.0	90.0	110.0	120.0
0.025	h_{sx}	62.5	75.0	87.5	100.0	112.5	137.5	150.0

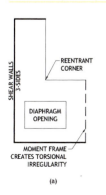
(a)

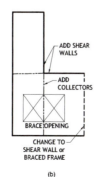
(b)

Figure 7.10 (a) Common horizontal seismic irregularities and (b) their mitigation

Timber Lateral Design

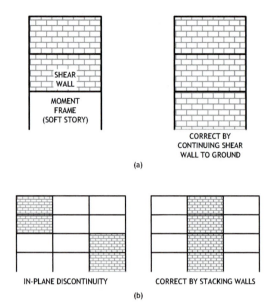

Figure 7.11 Vertical seismic irregularities showing (a) soft story and (b) in-plane discontinuities, and their mitigation

on three sides, with a moment frame on the fourth, illustrated in Figure 7.10;
- Reentrant Corners, which occur where there is an inside corner of the structure without frames or shear walls along them, shown also in Figure 7.10;
- Diaphragm Discontinuity, which happens where there are large openings in the diaphragms;
- Out-of-Plane Offsets, which occur where the lateral system changes plane and the forces must be transferred through the diaphragm to the vertical frames or shear walls.

Vertical structural irregularities include:

- Soft Stories, which occur where there is a drastic change in stiffness between levels. For example, a shear wall sitting on a moment frame, shown in Figure 7.11;
- Weak Stories, which exist where there is a large change in strength between levels;

- Mass irregularities, which occur where the adjacent story is 50% heavier than the adjacent stories;
- Geometric irregularities, which happen when the horizontal dimension of the lateral system changes more than 30% longer than an adjacent story;
- In-Plane Discontinuities, which occur where the lateral system changes locations horizontally, creating overturning forces in the members below, illustrated in Figure 7.11.

Each irregularity comes with specific, and sometimes exhaustive code requirements. Some of them are not permitted for seismic design categories D through F. Any lateral system with these irregularities will have financial and environmental costs, as these systems always require more material to carry the required loads. Additionally, no amount of analysis or detailing will make these structures perform as well as buildings without them.

7.5.4 Seismic Force Amplifications

There are a handful of cases where the code requires that the basic seismic forces be amplified. These include low redundancy conditions, structural irregularities, and protection of certain lateral frame elements.

Redundancy is the ability of a structure to sustain damage without becoming unstable. If failure of a key element of the lateral force resisting system results in a reduction of story shear strength of greater than 33%, the seismic force must be increased by 30%. More, smaller frames or walls will result in a less expensive, better performing seismic system.

When we have structural irregularities, as discussed in the previous section, we must increase the forces in elements that will be affected by these. The amount depends upon the irregularity and member, but ranges from 25% to 300%.

7.5.5 Material Requirements

Masonry in seismic regions must utilize Type S or M mortar.

7.5.6 Shear Wall Reinforcing

To ensure shear walls perform adequately during an earthquake, and absorb energy, TMS 402 requires certain reinforcing and detailing

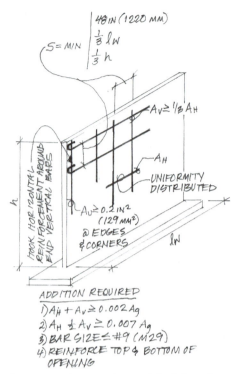

Figure 7.12 Reinforcing requirements for special reinforced masonry shear walls

requirements. These are shown in Figure 7.12, and center around limiting the spacing of reinforcing and ensuring there is enough rebar in the wall.

In cases where shear walls have high axial loads, the wall ends are prone to crushing. In these cases, we add boundary elements, like that in Figure 7.13.

We also limit the boundary reinforcing steel, so it yields well before the masonry crushes, as discussed in Section 2.4.3. Using equation 2.7, and $\alpha = 4.0$ for special reinforced masonry shear walls, we can calculate the maximum reinforcing steel A_{max}. Table 7.3 provides these values reinforcing for various combinations of bd.

To ensure adequate energy absorption, we need to ensure the wall will yield in bending, before its shear strength is exceeded. We do this by

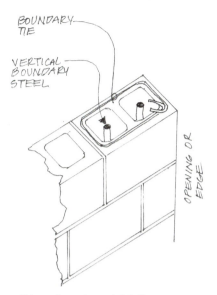

Figure 7.13 Shear wall boundary element detail

ensuring the shear strength is 125% higher than the shear that causes the wall to yield. In equation form, the shear demand V_{uY} is given by:

$$V_{uY} = \frac{1.25 M_u}{h/n} \tag{7.4}$$

where:

h = wall height, ft (m)
n = 1 for cantilevered walls, 2 for piers in double curvature

7.5.7 Non-Participating Seismic Elements

Members that are not part of the lateral seismic system move with the building. This can induce deformations greater than what they are designed for. TMS 402 requires additional reinforcing to ensure these elements will not fail during an earthquake.

7.6 WHERE WE GO FROM HERE

This chapter has introduced the general concepts of lateral design. From here, we estimate lateral forces on a structure, and determine their

Table 7.3 Maximum reinforcing area in special reinforced masonry shear walls

bd (in^2)	Maximum Reinforcing Area A_{max}, in^2						
	Masonry Compressive Strength (lb/in^2)						
	1,700	1,900	2,000	2,250	2,500	2,750	3,000
ρ_{max}	0.00421	0.00470	0.00495	0.00557	0.00619	0.00681	0.00742
46	0.194	0.216	0.228	0.256	0.285	0.331	0.342
64	0.269	0.301	0.317	0.356	0.396	0.436	0.475
80	0.337	0.376	0.396	0.445	0.495	0.544	0.594
96	0.404	0.451	0.475	0.535	0.594	0.653	0.713
120	0.505	0.564	0.594	0.668	0.742	0.817	0.891
144	0.606	0.677	0.713	0.802	0.891	0.980	1.07
160	0.673	0.752	0.792	0.891	0.990	1.09	1.19
192	0.808	0.903	0.950	1.07	1.19	1.31	1.43
240	1.01	1.13	1.19	1.34	1.48	1.63	1.78
288	1.21	1.35	1.43	1.60	1.78	1.96	2.14
320	1.35	1.50	1.58	1.78	1.98	2.18	2.38
384	1.62	1.81	1.90	2.14	2.38	2.61	2.85
448	1.88	2.11	2.22	2.49	2.77	3.05	3.33
480	2.02	2.26	2.38	2.67	2.97	3.27	3.56
560	2.36	2.63	2.77	3.12	3.46	3.81	4.16
672	2.83	3.16	3.33	3.74	4.16	4.57	4.99
768	3.23	3.61	3.80	4.28	4.75	5.23	5.70
864	3.63	4.06	4.28	4.81	5.35	5.88	6.41

1) Values for concrete masonry units

distribution into diaphragms and shear walls. This yields the shear and bending moment to select wall reinforcing. We then detail the structure, paying particular attention to the seismic requirements discussed above.

Seismic lateral design has become increasingly sophisticated in the past two decades. Prescriptive code requirements are giving way to performance-based design (PBD). This allows the owner and designer to pair the earthquake magnitude and structural performance that is consistent with the function of the building. Additionally, engineers are using PBD for more traditional, code-based buildings to reduce material consumption, as discussed in the *Special Topics* volume of this series.

Masonry Anchorage

Chapter 8

Paul W. McMullin

8.1 Anchor Types
8.2 Failure Modes
8.3 Capacity
8.4 Demand versus Capacity
8.5 Detailing Consideration
8.6 Anchorage Example
8.7 Where We Go from Here

Anchorage provides the connective interface between masonry and materials, such as steel and timber. Common applications include the connection of wood top plates and ledgers, **headed stud anchors** on embedded plates to make a welded connection illustrated in Figure 8.1, and anchors for veneer support angles. Anchors consist of two parts: internal and surface elements. Internal components engage the masonry and include anchor bolts, headed studs, or drilled anchors. Surface components consist of plates, angles, and nuts that complete the connection.

We use Section 9.1.6 of *Building Code Requirements for Masonry Structures*, TMS 402[1] for strength design of masonry anchorage. This covers the vast majority of conditions, but does not consider fatigue due to vibrating equipment, traffic loads, impact, or blast loads. Our goal for this chapter is to become familiar with design requirements and gain an understanding of what variables influence anchorage design.

8.1 ANCHOR TYPES

Anchors are either cast-in-place or post-installed, illustrated in Figure 8.2. Cast anchors commonly consist of headed anchor bolts and headed stud anchors (HSA), illustrated in Figure 8.3. They are tied in place before the grout is cast. Conversely, post-installed anchors are placed after the concrete is hard. They may be required to compensate

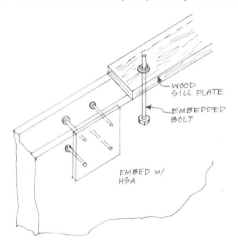

Figure 8.1 Common masonry anchorage configurations

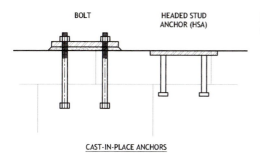

Figure 8.2 Cast-in-place and post-installed anchors

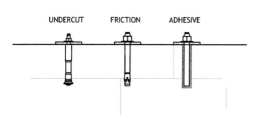

Figure 8.3 Headed anchor bolts and headed stud anchors prior to installation

for installation or construction errors, for ease of installation, or to retrofit an existing condition. Figure 8.4 shows a variety of post-installed anchors.

8.1.1 Cast-in-Place

Bolts and headed studs are common cast-in-place anchors. Bolts facilitate the attachment of steel plate or wood members. Headed studs

Masonry Anchorage

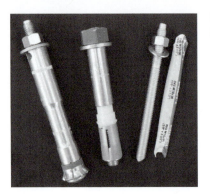

Figure 8.4 Post-installed anchors prior to installation

are welded to the plate and range from ⅜ in (9.5 mm) to 1 in (25 mm) diameter. Masons notch the plate into the masonry block, with the studs extending into the grout space. Studs and plate are tied in place before the grout is placed, shown in Figure 8.5. Because these connectors are built into the masonry, we get an integral, robust connection.

Anchor bolts that comply with ASTM F1554 come in three strength grades: 36 k/in² (245 MN/m²), 55 k/in² (375 MN/m²), and 105 k/in² (720 MN/m²)—the first two being preferred. Their diameters range from ¼ in (6 mm) to 4 in (100 mm) and can be fabricated to virtually any length. Mild steel headed stud anchors comply with ASTM A108. Table 8.1 shows

Figure 8.5 Embed plate and HSA prior to grouting

Table 8.1 Headed stud anchor dimensions

Imperial			
Shank		Head	
Diameter d_b (in)	Area A_b (in²)	Diameter d_{hs} (in)	Thickness t_{hs} (in)
1/4	0.0491	1/2	3/16
3/8	0.110	3/4	9/32
1/2	0.196	1	9/32
5/8	0.307	1 1/4	9/32
3/4	0.442	1 1/4	3/8
7/8	0.601	1 3/8	3/8
1	0.785	1 5/8	3/8
Metric			
Shank		Head	
Diameter d_b (mm)	Area A_b (mm²)	Diameter d_{hs} (mm)	Thickness t_{hs} (mm)
6.4	32	12.7	4.8
9.5	71	19.1	7.1
12.7	127	25.4	7.1
15.9	198	31.8	7.1
19.1	285	31.8	9.5
22.2	388	34.9	9.5
25.4	507	41.3	9.5

Source: PCI Design Handbook, 7th Edition

Table 8.2 Mechanical properties of anchor material

Imperial		
Material	*Yield (f_y)*	*Ultimate (f_u)*
	(k/in^2)	(k/in^2)
HSA	51	65
DBA	70	80
F1554 Gr 36	36	58
F1554 Gr 55	55	75
F1554 Gr 55	105	125
A307 Gr C	36	58
A36	36	58
Metric		
Material	*Yield (f_y)*	*Ultimate (f_u)*
	(MN/m^2)	(MN/m^2)
HSA	350	450
DBA	480	550
F1554 Gr 36	250	400
F1554 Gr 55	380	520
F1554 Gr 55	725	860
A307 Gr C	250	400
A36	250	400

the geometry of commonly used headed studs. They come in lengths from 3 to 10 in (75–255 mm). Table 8.2 summarizes the mechanical properties of anchor rods and headed studs.

8.1.2 Post-Installed

Post-installed anchors fall into two categories; mechanical and adhesive. **Mechanical anchors** transfer the loads to masonry by mechanical interlock or friction between the steel, and block, brick and grout, shown in Figure 8.5. **Adhesive anchors** transfer the load by chemical bond. Figure 8.6a shows the anchor components before installation, and Figure 8.6b shows a cutaway view of an installed reinforcing rod. The adhesive may be an epoxy, acrylic, or a hybrid. Post-installed anchors

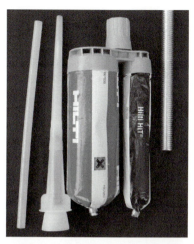

Figure 8.6 Adhesive anchor (a) before and (b) after installation (cutaway)

can be placed after the grout is seven days old and loaded once it has reached its design compressive strength.

Testing is fundamental to post-installed anchor performance. Require testing after the contractor has placed the post-installed anchors. This identifies installation problems and motivates the contractor. Base test forces on the following:

- Adhesive: the lesser of 50% of the ultimate bond strength and 80% of the steel yield strength
- Mechanical: 200% of allowable design values.

For common anchors, these general requirements translate into the values provided in Table 8.3.

8.2 FAILURE MODES

Anchorage failure modes consider the strength of the steel and masonry, with two major divisions: tension and shear. Tension failure modes are

Masonry Anchorage

Table 8.3 Post-installed anchor testing values in grouted masonry

Imperial			Metric		
Adhesive Anchors			**Adhesive Anchors**		
Anchor Diameter	*Embedment*	*Test Force*	*Anchor Diameter*	*Embedment*	*Test Force*
(in)	*(in)*	*(lb)*	*(mm)*	*(mm)*	*(kN)*
3/8	3 3/8	2,500	10	86	11,130
1/2	4 1/2	3,500	13	114	15,570
5/8	5 5/8	3,800	16	143	16,910
3/4	6 3/4	4,000	19	171	17,800
Mechanical Anchors			**Mechanical Anchors**		
Anchor Diameter	*Embedment*	*Test Force*	*Anchor Diameter*	*Embedment*	*Test Force*
(in)	*(in)*	*(lb)*	*(in)*	*(in)*	*(kN)*
3/8	2 5/8	870	10	67	3,870
1/2	3 1/2	1,100	13	89	4,900
5/8	4 3/8	1,800	16	111	8,010
3/4	5 1/4	2,100	19	133	9,350

Source: Ingenium Design

steel yielding and masonry breakout shown in Figure 8.7. Shear failure modes include steel yield, masonry breakout, crushing, and **pryout**, illustrated in Figure 8.8.

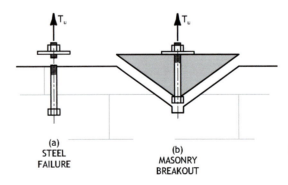

(a) STEEL FAILURE
(b) MASONRY BREAKOUT

Figure 8.7 Tension failure modes in masonry anchorage

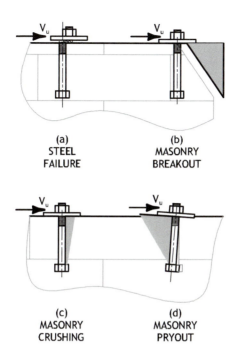

Figure 8.8 Shear failure modes in masonry anchorage

Let's now get into designing anchorage for these failure modes.

8.3 CAPACITY

The masonry code provides requirements for strength reduction factors, steel and concrete strength, **anchor group** effects, and seismic design. Strength reduction factors ϕ are 0.5 for masonry failure modes, and 0.9 for steel modes.

8.3.1 Tension

8.3.1.a Steel Strength

We calculate the tension strength B_{ans} of steel anchors as follows:

$$B_{ans} = A_b f_y \tag{8.1}$$

Masonry Anchorage

where:

A_b = anchor cross-sectional area, in² (mm²)
f_y = steel yield strength, k/in² (MN/m²)

Using equation (8.1), Table 8.4 shows the steel strength of headed stud anchors and bolts of various diameters. This typically represents the upper bound strength of a masonry anchor.

Table 8.4 Bolt and HSA steel strength

Imperial

Shank Diameter	Section Area	Tension ϕB_{ans}		Shear ϕB_{anv}	
d_b	A_b	Bolt	HSA	Bolt	HSA
(in)	(in²)	(k)	(k)	(k)	(k)
¼	0.0491	1.59	2.25	0.954	1.35
⅜	0.110	3.58	5.07	2.15	3.04
½	0.196	6.36	9.01	3.82	5.41
⅝	0.307	9.94	14.1	5.96	8.45
¾	0.442	14.3	20.3	8.59	12.2
⅞	0.601	19.5	27.6	11.7	16.6
1	0.785	25.4	36.0	15.3	21.6

Metric

Shank Diameter	Section Area	Tension ϕB_{ans}		Shear ϕB_{anv}	
d_b	A_b	Bolt	HSA	Bolt	HSA
(mm)	(mm²)	(kN)	(kN)	(kN)	(kN)
6.35	31.7	7.13	9.98	4.28	5.99
9.53	71.3	16.0	22.4	9.62	13.5
12.7	127	28.5	39.9	17.1	23.9
15.9	198	44.5	62.3	26.7	37.4
19.1	285	64.1	89.8	38.5	53.9
22.2	388	87.3	122	52.4	73.3
25.4	507	114	160	68.4	95.8

8.3.1.b Masonry Breakout Strength

Masonry breakout is generally the most critical failure mode in tension. It particularly matters when the anchors are close to the member edge or have shallow **embedment**.

For headed anchors, masonry **breakout strength** in tension B_{anb} is given as:

(8.2)

$$B_{anb} = 4A_{pt}\sqrt{f'_m} \qquad B_{anb} = 0.332 A_{pt}\sqrt{f'_m}$$

where:

A_{pt} = projected tension area on masonry surface of a right circular cone, in^2 (mm^2)

$$A_{pt} = \min \begin{vmatrix} \pi l_b^2 \\ \pi l_e^2 \end{vmatrix}$$

l_b = embedment length, in (mm)
l_e = edge distance, in (mm)
f'_m = masonry compressive strength, lb/in^2 (MN/m^2)

The units of B_{anb} are *lb* for Imperial units, and *N* for Metric.

The projected tension area A_{pt} is critical to understand and must be modified when it intersects a free edge. Where the embedment length l_b is greater than the edge distance l_e, as in Figure 8.9a, we use the first equation for A_{pt}. We have no reduction in strength. Where the edge distance is less than the embedment, we must reduce the failure cone, as illustrated in Figure 8.9b, and use the second equation. This uses the edge distance as the radius for the area at the surface.

8.3.2 Shear

Shear strength is highly affected by edge distance. This is less pronounced when the shear force is parallel to a free edge, but critical when perpendicular to the edge, illustrated in Figure 8.10. Anchor shear strength equations are like tension but include two more masonry failure modes.

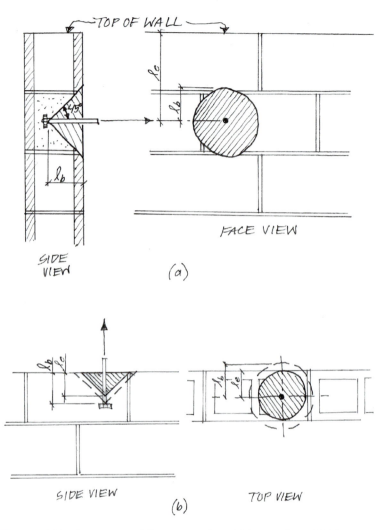

Figure 8.9 Tension masonry failure cones where (a) l_b controls, and (b) l_e controls

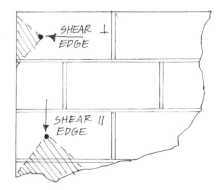

Figure 8.10 Shear in a masonry anchor (a) parallel, and (b) perpendicular to an edge

8.3.2.a Steel Strength

The shear strength B_{ans} of steel anchors is as follows:

$$B_{vns} = 0.6 A_b f_y \tag{8.3}$$

where:

A_b = anchor cross-sectional area, in² (mm²)
f_y = steel yield strength, k/in² (MN/m²)

The 0.6 accounts for the shear yield strength being 60% of the tension yield strength.

8.3.2.b Masonry Breakout Strength

Breakout shear capacity governs when close to an edge. It is given by:

$$\tag{8.4}$$

$$B_{vnb} = 4 A_{pv} \sqrt{f'_m} \qquad\qquad B_{vnb} = 0.332 A_{pv} \sqrt{f'_m}$$

where:

A_{pv} = projected shear area on masonry surface of half a right circular cone, in² (mm²), see Figure 8.11

$$A_{pv} = \frac{\pi l_e^2}{2}$$

l_e = edge distance, in (mm)
f'_m = masonry compressive strength, lb/in² (MN/m²)

Masonry Anchorage

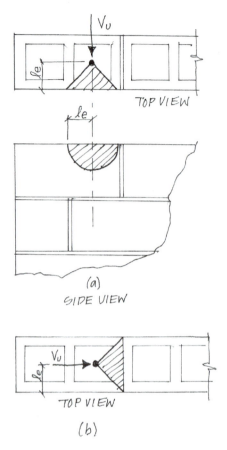

Figure 8.11 Shear masonry failure cones with the force (a) perpendicular, and (b) parallel to the top of the wall

The units of B_{vnb} are *lb* for Imperial units, and *N* for Metric.

Looking at the projected shear area for the top of a masonry wall, Figure 8.11 shows the force perpendicular and parallel to the wall. For both cases the edge distance l_e is the same. However, when the force is parallel to the wall, shown in Figure 8.11b, there is masonry restraining the anchor, which will certainly give a higher capacity. In this case, it is reasonable to double or triple the shear area A_{pv}.

8.3.2.c Masonry Crushing Strength

In cases where the free edge is quite far from the anchor, breakout no longer controls. In this case, it is likely that the masonry will crush at the steel interface, before the anchor steel yields. We calculate crushing strength as follows:

(8.5)

$$B_{vnc} = 1050 \sqrt[4]{f'_m A_b} \qquad B_{vnc} = 3216 \sqrt[4]{f'_m A_b}$$

where:

A_b = anchor cross-sectional area, in² (mm²)
f'_m = masonry compressive strength, lb/in² (MN/m²)

8.3.2.d Masonry Pryout Strength

Masonry pryout occurs when short headed anchors are loaded in shear. Since the studs are short and stiff, they push the concrete up perpendicular to the shear load (see Figure 8.8). Anchor pryout strength is calculated as follows:

(8.6)

$$B_{vnpry} = 2B_{anb} = 8A_{pt}\sqrt{f'_m} \qquad B_{vnpry} = 2B_{anb} = 0.664 A_{pt}\sqrt{f'_m}$$

BOX 8.1 INITIAL ANCHOR SIZING

A simple starting point for masonry anchorage is tension strength of around 4 k (18 kN) and shear capacity about 2 k (9 kN). Using these will give you a reasonable starting point.

Taking this to the next step, Table 8.5 through Table 8.7 show tension capacities for single, double, and quadruple anchor configurations, for a variety of masonry strength, embedment depth, and edge distance. It is necessary to also check the steel strength, given in Table 8.4. Similarly, Table 8.8 to Table 8.10 provide shear capacities for single, double, and quadruple anchor configurations, for various masonry strength, embedment depth, and bolt diameters. They cover a wide range of situations, and work for both the top and face of a wall installation.

Table 8.5 Masonry tension strength of a single anchor

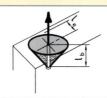

Compressive Strength f'_m (lb/in^2)	Embedment Depth l_b (in)	Tension Capacity ϕB_{anb} (k) Edge Distance l_e (in)					
		2¾	3¾	4¾	5¾	7¾	12
1,700	4	1.96	3.64	4.14	4.14	4.14	4.14
	5	1.96	3.64	5.85	6.48	6.48	6.48
	7	1.96	3.64	5.85	8.57	**12.7**	**12.7**
	9	1.96	3.64	5.85	8.57	**15.6**	**21.0**
	12	1.96	3.64	5.85	8.57	**15.6**	**37.3**
2,000	4	2.13	3.95	4.50	4.50	4.50	4.50
	5	2.13	3.95	6.34	7.02	7.02	7.02
	7	2.13	3.95	6.34	9.29	**13.8**	**13.8**
	9	2.13	3.95	6.34	9.29	**16.9**	**22.8**
	12	2.13	3.95	6.34	9.29	**16.9**	**40.5**
2,500	4	2.38	4.42	5.03	5.03	5.03	5.03
	5	2.38	4.42	7.09	7.85	7.85	7.85
	7	2.38	4.42	7.09	**10.4**	**15.4**	**15.4**
	9	2.38	4.42	7.09	**10.4**	**18.9**	**25.4**
	12	2.38	4.42	7.09	**10.4**	**18.9**	**45.2**

Table 8.5 continued

Compressive Strength f'_m (MN/mm²)	Embedment Depth l_b (mm)	Tension Capacity ϕB_{anb} (kN) Edge Distance l_e (mm)					
		70	95	120	145	200	305
12	100	8.71	16.2	18.4	18.4	18.4	18.4
	130	8.71	16.2	26.0	28.8	28.8	28.8
	180	8.71	16.2	26.0	38.1	**56.5**	**56.5**
	230	8.71	16.2	26.0	38.1	**69.2**	**93.3**
	300	8.71	16.2	26.0	38.1	**69.2**	**166**
14	100	9.45	17.6	20.0	20.0	20.0	20.0
	130	9.45	17.6	28.2	31.2	31.2	31.2
	180	9.45	17.6	28.2	41.3	**61.2**	**61.2**
	230	9.45	17.6	28.2	41.3	**75.1**	**101**
	300	9.45	17.6	28.2	41.3	**75.1**	**180**
17	100	10.6	19.7	22.4	22.4	22.4	22.4
	130	10.6	19.7	31.5	34.9	34.9	34.9
	180	10.6	19.7	31.5	**46.2**	**68.5**	**68.5**
	230	10.6	19.7	31.5	**46.2**	**83.9**	**113**
	300	10.6	19.7	31.5	**46.2**	**83.9**	**201**

1) This table only presents masonry breakout strength. Use Table 8.4 to compare with steel strength. Bold values indicate where steel controls for 5/8 in (16 mm) dia anchors.

2) This table is not for bent bar anchors.

Table 8.6 Masonry tension strength of two anchors

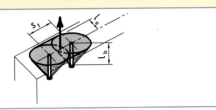

Compressive Strength f'_m (lb/in²)	Embedment Depth l_b (in)	Tension Capacity ϕB_{anb} (k) Edge Distance l_e (in)					
		2 ¾	3 ¾	4 ¾	5 ¾	7 ¾	12
1,700	4	3.92	7.29	8.10	8.10	8.10	8.10
	5	3.92	7.29	10.5	11.4	11.4	11.4
	7	3.92	7.29	10.5	14.3	19.6	19.6
	9	3.92	7.29	10.5	14.3	**23.2**	**29.9**
	12	3.92	7.29	10.5	14.3	23.2	49.2
2,000	4	4.25	7.90	8.79	8.79	8.79	8.79
	5	4.25	7.90	11.4	12.4	12.4	12.4
	7	4.25	7.90	11.4	15.5	**21.3**	**21.3**
	9	4.25	7.90	11.4	15.5	**25.2**	**32.4**
	12	4.25	7.90	11.4	15.5	25.2	53.3
2,500	4	4.75	8.84	9.83	9.83	9.83	9.83
	5	4.75	8.84	12.8	13.9	13.9	13.9
	7	4.75	8.84	12.8	17.3	**23.8**	**23.8**
	9	4.75	8.84	12.8	17.3	**28.2**	**36.2**
	12	4.75	8.84	12.8	17.3	**28.2**	59.6

Table 8.6 continued

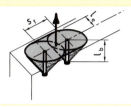

Compressive Strength f'_m (MN/mm²)	Embedment Depth l_b (mm)	Tension Capacity ϕB_{anb} (kN) Edge Distance l_e (mm)					
		70	95	120	145	200	305
12	100	17.4	32.4	36.0	36.0	36.0	36.0
	130	17.4	32.4	46.9	50.8	50.8	50.8
	180	17.4	32.4	46.9	63.4	87.3	87.3
	230	17.4	32.4	46.9	63.4	**103**	**133**
	300	17.4	32.4	46.9	63.4	**103**	**219**
14	100	18.9	35.2	39.1	39.1	39.1	39.1
	130	18.9	35.2	50.9	55.1	55.1	55.1
	180	18.9	35.2	50.9	68.8	**94.7**	**94.7**
	230	18.9	35.2	50.9	68.8	**112**	**144**
	300	18.9	35.2	50.9	68.8	**112**	**237**
17	100	21.1	39.3	43.7	43.7	43.7	43.7
	130	21.1	39.3	56.9	61.6	61.6	61.6
	180	21.1	39.3	56.9	76.9	**106**	**106**
	230	21.1	39.3	56.9	76.9	**125**	**161**
	300	21.1	39.3	56.9	76.9	**125**	**265**

1) This table only presents masonry breakout strength. Use Table 8.4 to compare with steel strength. Bold values indicate where steel controls for 5/8 in (16 mm) dia anchors.

2) This table is not for bent bar anchors.

3) s_1 = 6 in (150 mm)

Table 8.7 Masonry tension strength of four anchors

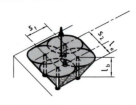

Compressive Strength f'_m (lb/in²)	Embedment Depth l_b (in)	Tension Capacity ϕB_{anb} (k) Edge Distance l_e (in)					
		2 ¾	3 ¾	4 ¾	5 ¾	7 ¾	12
1,700	4	7.84	14.0	15.0	15.0	15.0	15.0
	5	7.84	14.0	18.2	19.3	19.3	19.3
	7	7.84	14.0	18.2	22.9	29.5	29.5
	9	7.84	14.0	18.2	22.9	33.9	**41.8**
	12	7.84	14.0	18.2	22.9	33.9	**64.0**
2,000	4	8.50	15.2	16.3	16.3	16.3	16.3
	5	8.50	15.2	19.8	21.0	21.0	21.0
	7	8.50	15.2	19.8	24.9	32.0	32.0
	9	8.50	15.2	19.8	24.9	36.7	**45.3**
	12	8.50	15.2	19.8	24.9	36.7	**69.4**
2,500	4	9.50	17.0	18.2	18.2	18.2	18.2
	5	9.50	17.0	22.1	23.5	23.5	23.5
	7	9.50	17.0	22.1	27.8	35.8	35.8
	9	9.50	17.0	22.1	27.8	**41.1**	**50.6**
	12	9.50	17.0	22.1	27.8	**41.1**	**77.6**

Table 8.7 continued

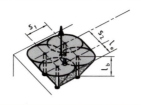

Compressive Strength f'_m (MN/mm²)	Embedment Depth l_b (mm)	Tension Capacity ϕB_{anb} (kN) Edge Distance l_e (mm)					
		70	95	120	145	200	305
12	100	34.9	62.4	66.9	66.9	66.9	66.9
	130	34.9	62.4	81.0	86.0	86.0	86.0
	180	34.9	62.4	81.0	101.9	131.3	131
	230	34.9	62.4	81.0	101.9	150.6	**186**
	300	34.9	62.4	81.0	101.9	150.6	**285**
14	100	37.8	67.7	72.5	72.5	72.5	72.5
	130	37.8	67.7	87.9	93.3	93.3	93.3
	180	37.8	67.7	87.9	110.6	142.4	142.4
	230	37.8	67.7	87.9	110.6	163.4	**202**
	300	37.8	67.7	87.9	110.6	163.4	**309**
17	100	42.3	75.7	81.1	81.1	81.1	81.1
	130	42.3	75.7	98.3	104.3	104.3	104.3
	180	42.3	75.7	98.3	123.6	159.2	159.2
	230	42.3	75.7	98.3	123.6	**183**	**225**
	300	42.3	75.7	98.3	123.6	**183**	**345**

1) This table only presents masonry breakout strength. Use Table 8.4 to compare with steel strength. Bold values indicate where steel controls for 5/8 in (16 mm) dia anchors.
2) This table is not for bent bar anchors.
3) Do not use this table for top of wall connections.
4) $s_1 = s_2 = 6$ in (150 mm)

Table 8.8 **Masonry shear strength of a single anchor**

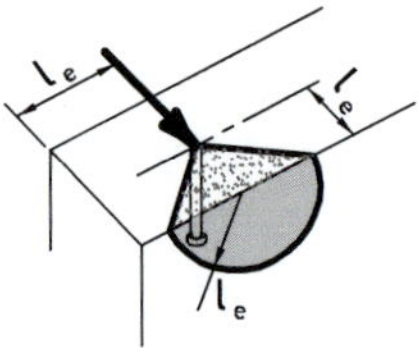

Compressive Strength f'_m *(lb/in²)*	*Bolt Diameter* d_b *(in)*	*Shear Capacity* ϕB_{vn} *(k)* *Edge Distance* l_e *(in)*					
		2 ¾	3 ¾	4 ¾	5 ¾	7 ¾	12
1,700	3/8	0.980	1.82	1.94	1.94	1.94	1.94
		0.980	1.82	2.24	2.24	2.24	2.24
	5/8	0.980	1.82	2.51	2.51	2.51	2.51
	3/4	0.980	1.82	2.75	2.75	2.75	2.75
	7/8	0.980	1.82	2.92	2.97	2.97	2.97
2,000	3/8	1.06	1.98	2.02	2.02	2.02	2.02
	1/2	1.06	1.98	2.34	2.34	2.34	2.34
	5/8	1.06	1.98	2.61	2.61	2.61	2.61
	3/4	1.06	1.98	2.86	2.86	2.86	2.86
	7/8	1.06	1.98	3.09	3.09	3.09	3.09
2,500	3/8	1.19	2.14	2.14	2.14	2.14	2.14
	1/2	1.19	2.21	2.47	2.47	2.47	2.47
	5/8	1.19	2.21	2.76	2.76	2.76	2.76
	3/4	1.19	2.21	3.03	3.03	3.03	3.03
	7/8	1.19	2.21	3.27	3.27	3.27	3.27

Table 8.8 *continued*

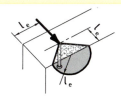

Compressive Strength f'_m (MN/mm²)	Bolt Diameter d_b (mm)	Shear Capacity ϕB_{vn} (kN) Edge Distance l_e (mm)					
		70	95	120	145	200	305
12	9.5	4.36	8.10	8.64	8.64	8.64	8.64
	12.7	4.36	8.10	9.98	9.98	9.98	9.98
	15.9	4.36	8.10	11.2	11.2	11.2	11.2
	19.1	4.36	8.10	12.2	12.2	12.2	12.2
	22.2	4.36	8.10	13.0	13.2	13.2	13.2
14	9.5	4.73	8.79	9.00	9.00	9.00	9.00
	12.7	4.73	8.79	10.4	10.4	10.4	10.4
	15.9	4.73	8.79	11.6	11.6	11.6	11.6
	19.1	4.73	8.79	12.7	12.7	12.7	12.7
	22.2	4.73	8.79	13.8	13.8	13.8	13.8
17	9.5	5.28	9.52	9.52	9.52	9.52	9.52
	12.7	5.28	9.83	11.0	11.0	11.0	11.0
	15.9	5.28	9.83	12.3	12.3	12.3	12.3
	19.1	5.28	9.83	13.5	13.5	13.5	13.5
	22.2	5.28	9.83	14.5	14.5	14.5	14.5

1) This table presents the minimum of masonry breakout, crushing, and pryout strength. Use Table 8.4 to compare with steel strength. All values in this table are less than the steel strength.

2) This table is not for bent bar anchors.

Table 8.9 Masonry shear strength of two anchors

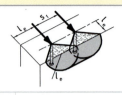

Compressive Strength f'_m (lb/in²)	Bolt Diameter d_b (in)	Shear Capacity ϕB_{vn} (k) Edge Distance l_e (in)					
		2¾	3¾	4¾	5¾	7¾	12
1,700	⅜	1.96	3.64	3.89	3.89	3.89	3.89
	½	1.96	3.64	4.49	4.49	4.49	4.49
	⅝	1.96	3.64	5.02	5.02	5.02	5.02
	¾	1.96	3.64	5.27	5.50	5.50	5.50
	⅞	1.96	3.64	5.27	5.94	5.94	5.94
2,000	⅜	2.13	3.95	4.05	4.05	4.05	4.05
	½	2.13	3.95	4.67	4.67	4.67	4.67
	⅝	2.13	3.95	5.23	5.23	5.23	5.23
	¾	2.13	3.95	5.72	5.72	5.72	5.72
	⅞	2.13	3.95	5.72	6.18	6.18	6.18
2,500	⅜	2.38	4.28	4.28	4.28	4.28	4.28
	½	2.38	4.42	4.94	4.94	4.94	4.94
	⅝	2.38	4.42	5.53	5.53	5.53	5.53
	¾	2.38	4.42	6.05	6.05	6.05	6.05
	⅞	2.38	4.42	6.39	6.54	6.54	6.54

Table 8.9 *continued*

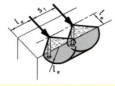

Compressive Strength f'_m (MN/mm²)	Bolt Diameter d_b (mm)	Shear Capacity ϕB_{vn} (kN) Edge Distance l_e (mm)					
		70	95	120	145	200	305
12	9.5	8.71	16.2	17.3	17.3	17.3	17.3
	12.7	8.71	16.2	20.0	20.0	20.0	20.0
	15.9	8.71	16.2	22.3	22.3	22.3	22.3
	19.1	8.71	16.2	23.5	24.5	24.5	24.5
	22.2	8.71	16.2	23.5	26.4	26.4	26.4
14	9.5	9.45	17.6	18.0	18.0	18.0	18.0
	12.7	9.45	17.6	20.8	20.8	20.8	20.8
	15.9	9.45	17.6	23.2	23.2	23.2	23.2
	19.1	9.45	17.6	25.4	25.5	25.5	25.5
	22.2	9.45	17.6	25.4	27.5	27.5	27.5
17	9.5	10.6	19.0	19.0	19.0	19.0	19.0
	12.7	10.6	19.7	22.0	22.0	22.0	22.0
	15.9	10.6	19.7	24.6	24.6	24.6	24.6
	19.1	10.6	19.7	26.9	26.9	26.9	26.9
	22.2	10.6	19.7	28.4	29.1	29.1	29.1

1) This table presents the minimum of masonry breakout, crushing, and pryout strength. Use Table 8.4 to compare with steel strength. All values in this table are less than the steel strength.
2) This table is not for bent bar anchors.
3) s_1 = 6 in (150 mm)

Table 8.10 Masonry shear strength of four anchors

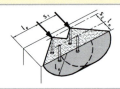

Compressive Strength f'_m (lb/in²)	Bolt Diameter d_b (in)	Shear Capacity ϕB_{vn} (k) Edge Distance l_e (in)					
		2 ¾	3 ¾	4 ¾	5 ¾	7 ¾	12
1,700	⅜	7.77	7.77	7.77	7.77	7.77	7.77
	½	8.98	8.98	8.98	8.98	8.98	8.98
	⅝	10.0	10.0	10.0	10.0	10.0	10.0
	¾	11.0	11.0	11.0	11.0	11.0	11.0
	⅞	11.9	11.9	11.9	11.9	11.9	11.9
2,000	⅜	8.10	8.10	8.10	8.10	8.10	8.10
	½	9.35	9.35	9.35	9.35	9.35	9.35
	⅝	10.5	10.5	10.5	10.5	10.5	10.5
	¾	11.4	11.4	11.4	11.4	11.4	11.4
	⅞	12.4	12.4	12.4	12.4	12.4	12.4
2,500	⅜	8.56	8.56	8.56	8.56	8.56	8.56
	½	9.88	9.88	9.88	9.88	9.88	9.88
	⅝	11.1	11.1	11.1	11.1	11.1	11.1
	¾	12.1	12.1	12.1	12.1	12.1	12.1
	⅞	13.1	13.1	13.1	13.1	13.1	13.1

Table 8.10 *continued*

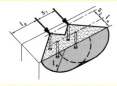

Compressive Strength f'_m (MN/mm^2)	Bolt Diameter d_b (mm)	Shear Capacity ϕB_{vn} (kN) Edge Distance l_e (mm)					
		70	95	120	145	200	305
12	9.5	34.6	34.6	34.6	34.6	34.6	34.6
	12.7	39.9	39.9	39.9	39.9	39.9	39.9
	15.9	44.6	44.6	44.6	44.6	44.6	44.6
	19.1	48.9	48.9	48.9	48.9	48.9	48.9
	22.2	52.8	52.8	52.8	52.8	52.8	52.8
14	9.5	36.0	36.0	36.0	36.0	36.0	36.0
	12.7	41.6	41.6	41.6	41.6	41.6	41.6
	15.9	46.5	46.5	46.5	46.5	46.5	46.5
	19.1	50.9	50.9	50.9	50.9	50.9	50.9
	22.2	55.0	55.0	55.0	55.0	55.0	55.0
17	9.5	38.1	38.1	38.1	38.1	38.1	38.1
	12.7	44.0	44.0	44.0	44.0	44.0	44.0
	15.9	49.2	49.2	49.2	49.2	49.2	49.2
	19.1	53.9	53.9	53.9	53.9	53.9	53.9
	22.2	58.2	58.2	58.2	58.2	58.2	58.2

1) This table presents the minimum of masonry breakout, crushing, and pryout strength. Use Table 8.4 to compare with steel strength. All values in this table are less than the steel strength.

2) This table is not for bent bar anchors.

3) Do not use this table for top of wall connections.

4) $s_1 = s_2 =$ 6 in (150 mm)

8.4 DEMAND VERSUS CAPACITY

In the previous section, we learned how to calculate tensile and shear capacity for individual and groups of anchors. For tension strength, we use the smallest steel, and concrete breakout strength and multiply it by the applicable strength reduction factor ϕ. For shear we use the smallest steel, concrete breakout, crushing, and pryout strength, again multiplied by ϕ. We then compare them with the tension and shear demand for that specific anchorage. The demand must always be smaller than the capacity. This relationship can be written as follows:

$$T_u \leq \phi B_{an} \tag{8.7}$$

$$V_u \leq \phi B_{vn}$$

where:

T_u = tension demand, k (kN)
V_u = shear demand, k (kN)
ϕB_{an} = tension capacity, k (kN)
ϕB_{vn} = shear capacity, k (kN)

When anchorage has both tension and shear loads, we must check the interaction effects using the following relationship. When it is less than one, our design is satisfactory.

$$\frac{T_u}{\phi B_{an}} + \frac{V_u}{\phi B_{vn}} \leq 1.0 \tag{8.8}$$

8.5 DETAILING CONSIDERATION

It is prudent to consider detailing as we design the anchorage group. Waiting to detail until the connection is fully designed will result in serious redesign time. Some things to keep in mind as you set up the connection geometry:

- Edge distance has the greatest effect on strength. The further you can get the anchor from the edge, the stronger the connection will be.
- Embedment depth has the next greatest effect on strength. Deeper is better.
- Eccentricity greatly increases the anchor forces and is often unnecessary. Keep the line of force through the anchor centroid to avoid eccentricity.

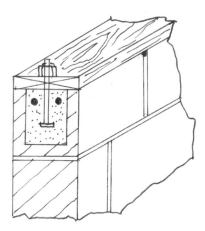

Figure 8.12 Supplemental anchorage reinforcing contributing connection reliability

- Use headed anchors, rather than 'J' or 'L' bolts, which have been shown to pull out under smaller loads.
- Provide supplemental reinforcing where possible, like that shown in Figure 8.12. It provides greater strength and reliability. Even though the masonry code doesn't have a way to include it in your calculations, it will make for a stronger connection.
- Ensure the anchor bolt or HSA does not interfere with the face shells, when oriented perpendicular to the walls.

BOX 8.2 DESIGN STEPS

Follow the steps below for anchorage design of cast-in-place anchors.
1. Draw the structural layout and anchorage geometry
2. Determine the shear and tensile loads acting on the anchorage (often the reaction at the end of a member)
3. Material parameters—Choose masonry compressive strength and anchor strength
4. Estimate initial size and number of anchors
5. Determine the anchor capacity, and check it against the demand
6. Summarize the results.

8.6 ANCHORAGE EXAMPLE

Step 1: Structural Layout

We will continue our maintenance shop example from Chapter 4, shown in Figure 4.8. We will change the roof members to deep I-joists and use a 3x wood ledger with anchors at 16 in (400 mm) on center, shown in Figure 8.13.

From our sketches, we get the following geometric data:

l_t = 19.33ft $\qquad l_t$ = 5.89m

s = 16in $\qquad s$ = 0.41m

h_w = 15ft $\qquad h_w$ = 4.57m

h_p = 2ft $\qquad h_p$ = 0.61m

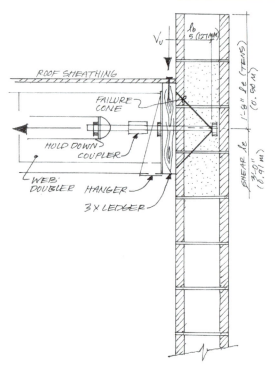

Figure 8.13 Example masonry anchor connection

Step 2: Determine the Loads

We need to find the vertical shear load V_u and horizontal seismic force as the wall moves away from the roof. For gravity loads, our roof dead and snow are as follows:

$$D_R = 20 \frac{lb}{ft^2} \qquad\qquad D_R = 0.96 \frac{kN}{m^2}$$

$$S = 50 \frac{lb}{ft^2} \qquad\qquad S = 2.39 \frac{kN}{m^2}$$

Using the snow heavy load combination, we get the vertical shear as follows:

$$V_u = (1.2 D_R + 1.6 S) l_t s$$

$$= \left[1.2\left(0.02 \frac{k}{ft^2}\right) + 1.6\left(0.05 \frac{k}{ft^2}\right)\right] 19.33 ft \left(16 in \frac{1 ft}{12 in}\right)$$
$$= 2.68 k$$

$$= \left[1.2\left(0.96 \frac{kN}{m^2}\right) + 1.6\left(2.39 \frac{kN}{m^2}\right) 5.89 m (0.41 m)\right]$$
$$= 12.0 kN$$

For seismic loads, we will also need to know the unit weight of the masonry lintel. Using an 8 in (203 mm) thick wall, normal-weight masonry, grouted at 24 in (600 mm) on center. From Table 2.3:

$$D_W = 55 \frac{lb}{ft^2} \qquad\qquad D_W = 2.63 \frac{kN}{m^2}$$

Using the wall and parapet heights, h_w and h_p, we get the weight tributary to the anchor.

$$W_a = D_W \left(\frac{h_w}{2} + h_p\right) s$$

$$= 0.55 \frac{k}{ft^2} \left(\frac{15 ft}{2} + 2 ft\right) 16 in \left(\frac{1 ft}{12 in}\right) \qquad = 2.63 \frac{kN}{m^2} \left(\frac{4.57 m}{2} + 0.61 m\right) 0.41 m$$
$$= 0.70 k \qquad\qquad\qquad\qquad\qquad\qquad = 3.12 kN$$

The out of plane seismic force is given as:

$$F_p = 0.8 S_{DS} I_e W_a$$

This equation is from ASCE 7, the structural load code. See Chapter 8 of *Introduction to Structures* for more discussion on seismic loads. We will take S_{DS} and I_e as follows:

$S_{DS} = 0.99$

$I_e = 1.0$

$F_p = 0.8(0.99)1.0(0.70k)$ $\qquad F_p = 0.8(0.99)1.0(3.12 kN)$
$\quad = 0.55k$ $\qquad\qquad\qquad\qquad = 2.47 kN$

Step 3: Masonry Parameters

We now need to determine our masonry and reinforcing strengths, and masonry modulus of elasticity. Using the values from previous examples the masonry compressive strength is:

$$f'_m = 1,700 \frac{lb}{in^2} \qquad\qquad f'_m = 11.72 \frac{MN}{m^2}$$

Using a low-grade bolt, we get the yield strength from Table 8.2.

$$f_y = 36 \frac{k}{in^2} \qquad\qquad f_y = 250 \frac{MN}{m^2}$$

Step 4: Initial Size

To get an initial anchor size, we go to Table 8.8 and see that ¾ in (19 mm) anchors cap out at 2.75 k (12.2 kN). Because our vertical shear is just below this, most likely our anchorage won't work the way we have it drawn. However, let's see how close it really is.

Step 5: Anchor Capacity

Shear Capacity

We now have what we need to calculate and check the strength. Starting with shear, we check the four failure modes: steel, breakout, crushing, and pryout.

We will need to identify the edge distance, and select an embedment depth. From Figure 8.13 we get:

$l_e = 36$ in $\qquad\qquad l_e = 910 mm$

$l_b = 5$ in $\qquad\qquad l_b = 127 mm$

which puts the head of the anchor near the inside of the face shell.

Using a ¾ in (19 mm) anchor, using Table 8.1 the area is:

$A_b = 0.442 \text{in}^2$ $\qquad\qquad A_b = 285 mm^2$

The steel strength is:

$$B_{vns} = 0.6 A_b f_y$$
$$= 0.6(0.442 in^2) 36 \frac{k}{in^2}$$
$$= 9.55 k$$

$$= 0.6(285 mm^2) 250,000 \frac{kN}{m^2} \left(\frac{1m}{1,000mm}\right)^2$$
$$= 42.8 kN$$

Masonry breakout strength is given as:

$$B_{vnb} = 4 A_{pv} \sqrt{f'_m} \qquad\qquad B_{vnb} = 0.332 A_{pv} \sqrt{f'_m}$$

The breakout area is given by:

$$A_{pv} = \frac{\pi l_e^2}{2}$$
$$= \frac{\pi (36 in)^2}{2}$$
$$= 2,036 in^2$$

$$= \frac{\pi (910 mm)^2}{2}$$
$$= 1.30 x 10^6 mm^2$$

$$B_{vnb} = 4(2,036 in^2)\sqrt{1,700 \frac{lb}{in^2}} \; \frac{1k}{1,000 lb}$$
$$= 336 k$$

$$B_{vnb} = 0.332(1.30 x 10^6 mm^2)\sqrt{11.72 \frac{MN}{m^2}} \; \frac{1kN}{1,000N}$$
$$= 1,478 kN$$

This is rather high, and is due to the long edge distance, and wont control. Next, we check masonry crushing strength, which likely will limit the capacity.

$$B_{vnc} = 1050 \frac{\sqrt[4]{f'_m A_b}}{1000}$$

$$= 1{,}050 \sqrt[4]{1{,}700 \frac{lb}{in^2}(0.442 in^2)} \frac{1k}{1{,}000 lb}$$

$$= 5.5k$$

$$B_{vnc} = 3216 \sqrt[4]{f'_m A_b}$$

$$B_{vnc} = 3.216 \sqrt[4]{11.72 \frac{MN}{m^2}(285 mm^2)} \frac{1kN}{1{,}000 N}$$

$$= 24.5 kN$$

And now for pryout, which is double the tensile breakout strength,

$$B_{vnpry} = 8 A_{pt} \sqrt{f'_m} \qquad\qquad B_{vnpry} = 0.664 A_{pt} \sqrt{f'_m}$$

To find the project tension failure surface A_{pt}, we take the least of l_b and l_e.

$$A_{pt} = \pi l_b^2$$

$$= \pi (5 in)^2 \qquad\qquad = \pi (127 mm)^2$$

$$= 78.6 in^2 \qquad\qquad = 50{,}671 mm^2$$

$$B_{vnpry} = 8(78.6 in^2)\sqrt{1{,}700 \frac{lb}{in^2}} \frac{1k}{1{,}000 lb}$$

$$= 25.9 k$$

$$B_{vnpry} = 0.664(50{,}671 mm^2)\sqrt{11.72 \frac{MN}{m^2}} \frac{1kN}{1{,}000 N}$$

$$= 115 kN$$

From this analysis, we see masonry crushing B_{vnc} controls. Applying the strength reduction factor, we get the anchor shear capacity as:

$$\phi = 0.5$$

$$\phi B_{vn} = \phi B_{vnc}$$

$$= 0.5 (5.5k) \qquad\qquad = 0.5 (24.5 kN)$$

$$= 2.75 k \qquad\qquad = 12.3 kN$$

This is barely larger than V_u, and works if we don't have a combined tension load. Let's look at the tension capacity and interaction formula first, before we decide what to do.

TENSION CAPACITY

The steel strength is:

$B_{ans} = A_b f_y$

$= 0.442 in^2 \left(36 \dfrac{k}{in^2} \right)$ $\quad = 285 mm^2 \left(250 \dfrac{MN}{m^2} \right) \left(\dfrac{1m}{1,000mm} \right)^2 \dfrac{1,000 kN}{1 MN}$

$= 15.9 k$ $\quad\quad\quad\quad\quad\quad\quad\quad = 71.3 kN$

Because the tension breakout strength is half the shear prying capacity, we have the variables we need to calculate tension breakout quickly.

$B_{anb} = 4 A_{pt} \sqrt{f'_m}$

$B_{anb} = 4 (78.6 in^2) \sqrt{1,700 \dfrac{lb}{in^2}} \dfrac{1k}{1,000 lb}$

$= 13.0 k$

$\quad\quad\quad\quad\quad B_{anb} = 0.332 A_{pt} \sqrt{f'_m}$

$\quad\quad\quad\quad\quad B_{anb} = 0.332 (50,671 mm^2) \sqrt{11.72 \dfrac{MN}{m^2}} \dfrac{1 kN}{1,000 N}$

$\quad\quad\quad\quad\quad = 57.6 kN$

It looks like masonry breakout controls over steel yield, so the tension capacity ϕB_{an} is:

$\phi B_{an} = \phi B_{anb}$

$= 0.5 (13.0 k)$ $\quad\quad\quad = 0.5 (57.6 kN)$

$= 6.5 k$ $\quad\quad\quad\quad\quad = 28.5 kN$

This is quite high compared to our demand, which is good. Let's now check the combined effect of tension and shear on our anchor, using the interaction equation 8.8.

$$\text{INT} = \frac{F_p}{\phi B_{an}} + \frac{V_u}{\phi B_{vn}}$$
$$= \frac{0.55k}{6.5k} + \frac{2.68k}{2.75} \qquad\qquad = \frac{2.47 kN}{28.5 kN} + \frac{12.0 kN}{12.3 kN}$$
$$= 1.06 \qquad\qquad\qquad\qquad\qquad = 1.06$$

Because this is larger than 1, our design doesn't work, barely.

Step 6: Summarize Results

A simple fix is to increase the bolt spacing to 12 in (300 mm) on center. However, this doesn't lay out well with the joist spacing. Let's choose an 8 in (200 mm) spacing. While it might seem excessive, it is always better to have reserve capacity in connections when possible. It also simplifies construction, which is always appreciated in the field.

To complete this design, we need to check the wood bearing strength at the bolt, and hold down capacity. These are discussed in *Timber Design* in this series.

8.7 WHERE WE GO FROM HERE

Many loads in masonry anchorages are reasonable and can be carried by one to four anchors. When tension loads become larger, and are parallel to the masonry element, we can utilize embed plates with deformed bar anchors (DBA) welded to them, shown in Figure 8.14. This allows the load to be transferred deep into the masonry. In this case, the design is driven by the DBA development length.

For large forces perpendicular to the masonry, we often figure out a way to reduce the load to the masonry, utilizing a channel to distribute high point loads to many anchors.

Understanding post-installed anchorage design is a good next step in furthering our capabilities. Generally, we obtain allowable shear and tension values from ICC Evaluation reports, and modify these to account for edge and spacing realities.

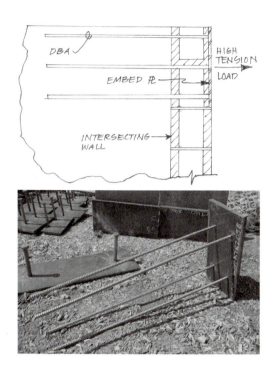

Figure 8.14 Embed plate with DBAs to carry high tension forces, (a) detail, (b) prior to installation

NOTE

1. TMS, *Building Code Requirements for Masonry Structures*, TMS 402–13 (Longmont, CO: The Masonry Society, 2013).

Appendix 1

Table A1.1 Masonry section properties (imperial)

Size	Vertical Grout Spacing	Net Section Properties (use for stress and strain)			Average Section Properties (use for stiffness and deflection)			
		A_n	I_n	S_n	A_{avg}	I_{avg}	S_{avg}	r_{avg}
	(in)	(in²/ft)	(in⁴/ft)	(in³/ft)	(in²/ft)	(in⁴/ft)	(in³/ft)	(in)
8 in Units	None	30.0	309	81.0	41.5	334	87.6	2.84
	16	62.0	379	99.3	65.8	387	102	2.43
	24	51.3	355	93.2	57.7	369	96.9	2.53
	32	46.0	344	90.1	53.7	361	94.6	2.59
	48	40.7	332	87.1	49.6	352	92.2	2.66
	Solid	91.5	443	116	91.5	443	116	2.20
10 in Units	None	30.0	530	110	48.0	606	126	3.55
	16	74.8	719	150	80.8	745	155	3.04
	24	59.8	656	136	69.9	699	145	3.16
	32	52.4	625	130	64.4	676	140	3.24
	48	44.9	593	123	58.9	652	136	3.33
	Solid	116	892	185	116	892	185	2.78
12 in Units	None	30.0	811	140	53.1	972	167	4.28
	16	87.3	1,209	208	95.0	1,262	217	3.64
	24	68.2	1,076	185	81.0	1,165	201	3.79
	32	58.7	1,010	174	74.1	1,117	192	3.88
	48	49.1	944	162	67.1	1,068	184	3.99
	Solid	140	1,571	270	140	1,571	270	3.36
16 in Units	None	30.0	1,554	199	63.2	2,031	260	5.67
	16	112	2,737	350	124	2,896	371	4.84
	24	85.0	2,343	300	103	2,608	334	5.02
	32	71.2	2,146	275	93.4	2,463	315	5.14
	48	57.5	1,948	249	83.3	2,319	297	5.28
	Solid	188	3,815	488	188	3,815	488	4.51

Source: NCMA TEK 14–1B

Table A1.1m Masonry section properties (metric)

Size	Vertical Grout Spacing (mm)	Net Section Properties (use for stress and strain)			Average Section Properties (use for stiffness and deflection)			
		$A_n \times 10^3$ (mm^2/m)	$I_n \times 10^6$ (mm^4/m)	$S_n \times 10^3$ (mm^3/m)	$A_{avg} \times 10^3$ (mm^2/m)	$I_{avg} \times 10^6$ (mm^4/m)	$S_{avg} \times 10^3$ (mm^3/m)	r_{avg} (mm)
203 mm Units	None	63.5	422	4,355	87.8	456	4,710	72.1
	406	131	517	5,339	139	529	5,457	61.7
	610	109	485	5,011	122	504	5,210	64.3
	813	97.4	469	4,844	114	492	5,086	65.8
	1,219	86.1	453	4,683	105	480	4,957	67.6
	Solid	194	605	6,253	194	605	6,253	55.9
254 mm Units	None	63.5	724	5,919	102	828	6,774	90.2
	406	158	982	8,038	171	1,017	8,317	77.2
	610	127	896	7,328	148	954	7,806	80.3
	813	111	853	6,978	136	922	7,548	82.3
	1,219	95.0	810	6,624	125	891	7,290	84.6
	Solid	244	1,218	9,962	244	1,218	9,962	70.6
305 mm Units	None	63.5	1,108	7,505	112	1,327	8,984	109
	406	185	1,651	11,183	201	1,724	11,677	92.5
	610	144	1,470	9,957	171	1,591	10,779	96.3
	813	124	1,379	9,344	157	1,525	10,333	98.6
	1,219	104	1,289	8,731	142	1,459	9,882	101
	Solid	295	2,145	14,532	295	2,145	14,532	85.3
406 mm Units	None	63.5	2,122	10,693	134	2,773	13,973	144
	406	238	3,738	18,839	261	3,955	19,930	123
	610	180	3,199	16,124	219	3,561	17,946	128
	813	151	2,930	14,763	198	3,364	16,951	131
	1,219	122	2,660	13,408	176	3,167	15,962	134
	Solid	397	5,209	26,252	397	5,209	26,252	115

Source: NCMA TEK 14–1B

Appendix 2

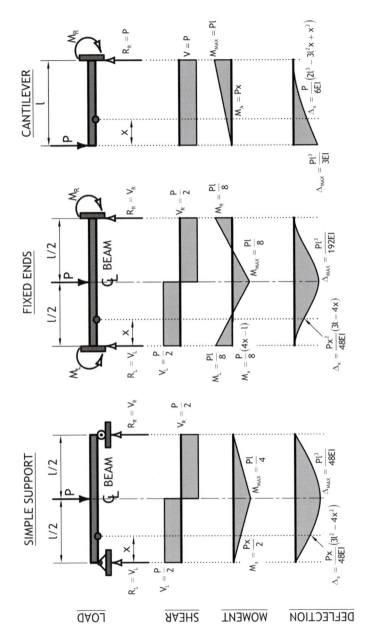

Figure A2.1 Point load, single-span, beam solutions and diagrams

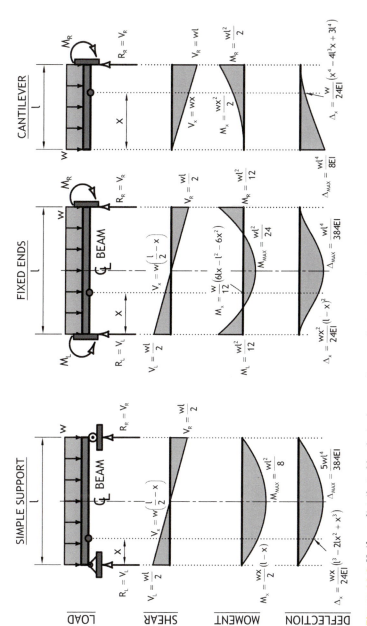

Figure A2.2 Uniform distributed load, single span, beam solutions and diagrams

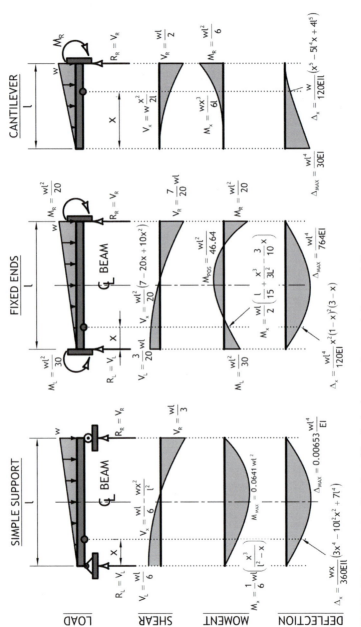

Figure A2.3 Triangular distributed load, single-span, beam solutions and diagrams

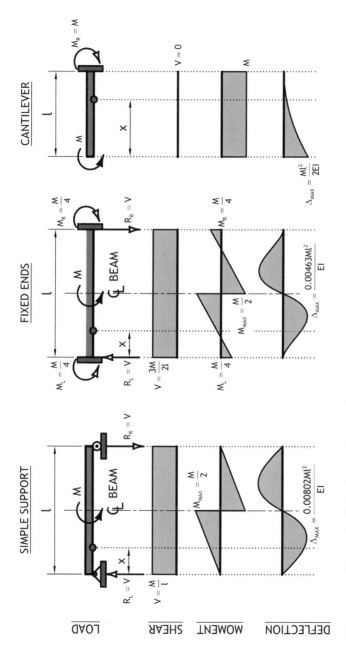

Figure A2.4 Moment load, single span, beam solutions and diagrams

Appendix 2

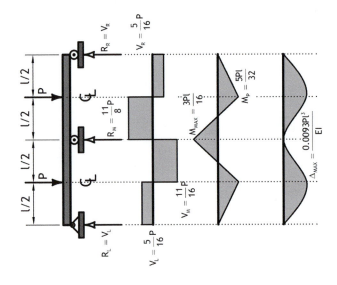
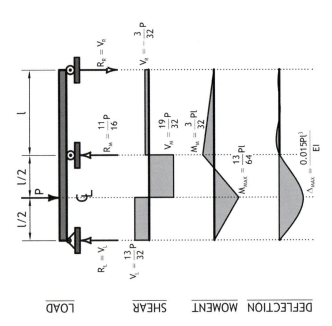

Figure A2.5 Point load, double-span, beam solutions and diagrams

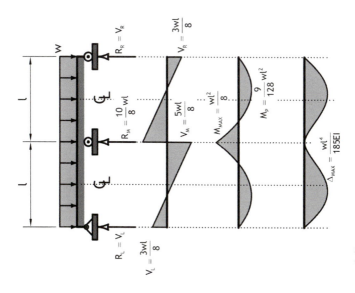
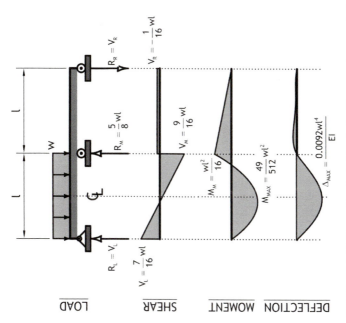

Figure A2.6 Uniform distributed load, double-span, beam solutions and diagrams

Appendix 3

Units		
Unit	*Definition*	*Typical Use*
°	degrees	angle
deg	degrees	angle
ft	feet	length
ft^2	square feet	area
ft^3	cubic feet	volume
hr	hour	time
in	inches	length
in^2	square inch	area
in^3	cubic inch	volume
in^4	inches to the fourth power	moment of inertia
k	kip (1000 pounds)	force
k/ft	kips per foot (aka klf)	distributed linear force
k/ft^2	kips per square foot (aka ksf)	distributed area force, pressure
k/ft^3	kips per cubic foot (aka kcf)	density
k/in^2	kips per square inch (aka ksi)	distributed area force, pressure
k-ft	kip-feet	moment, torque
lb, lb_f	pound	force
lb/ft	pounds per foot (aka plf)	distributed linear force
lb/ft^2	pounds per square foot (aka psf)	distributed area force, pressure
lb/ft^3	pounds per cubic foot (aka pcf)	density
lb/in^2	pounds per square inch (aka psi)	distributed area force, pressure
lb-ft	pound-feet	moment, torque
rad	radian	angle
yd^3	cubic yard	volume

Appendix 3

Units m

Unit	Definition	Typical Use
°	degrees	angle
deg	degrees	angle
g	gram	mass
hr	hour	time
kN	newton	force
kN	kiloNewton	force
kN/m	kiloNewton per meter	distributed linear force
kN/m^2	kiloNewton per square meter (aka kPa)	distributed area force, pressure
kN/m^3	kiloNewton per cubic foot	density
kN-m	kiloNewton-meter	moment, torque
m	meters	length
m^2	square meters	area
m^3	cubic meters	volume
min	minute	time
mm	millimeters	length
mm^2	square millimeters	area
mm^3	cubic millimeters	volume
mm^4	millimeters to the fourth power	moment of inertia
MN/m^2	meganewton per square meter (aka MPa)	distributed area force, pressure
N	newton	force
N/m	newtons per meter	distributed linear force
N/m^2	newtons per square meter (aka Pa)	distributed area force, pressure
N/m^3	newtons per cubic meter	density
N/mm^2	newtons per square millimeter (aka MPa)	distributed area force, pressure
Pa	newton per square meter (N/m^2)	distributed area force, pressure
rad	radian	angle

Appendix 4

Symbol	Definition	Imperial	Metric
$\#_b$	bending (flexure) property or action	vary	
$\#_c$	compression property or action	vary	
$\#_D$	dead load related action	vary	
$\#_H$	horizontal action	lb or k	N, kN, MN
$\#_L$	live load related action	vary	
$\#_L$	action on left	vary	
$\#_n$	nominal capacity	vary	
$\#_R$	action on right	vary	
$\#_S$	snow load related action	vary	
$\#_t$	tension property or action	vary	
$\#_u$	factored load, any type	vary	
$\#_v$	shear property or action	vary	
$\#_V$	vertical action	lb or k	N, kN, MN
$\#_W$	wind load related action	vary	
a	compression stress block depth	in	mm
A_b	bolt cross-section area	in^2	mm^2
A_g	gross cross-section area	in^2	mm^2
A_{max}	maximum area of reinforcing steel	in^2	mm^2
A_n	net cross section area	in^2	mm^2
A_{nv}	net shear area	in^2	mm^2
A_{pt}	projected tension area on masonry surface of right circular cone	in^2	mm^2
A_{pv}	projected shear area on masonry surface of one-half right circular cone	in^2	mm^2
A_s	reinforcing steel cross-section area	in^2	mm^2
A_v	shear reinforcing steel cross-section area	in^2	mm^2
A_v	masonry shear area in wall	in^2	mm^2
b	section width	in	mm

Symbol	Definition	**Imperial**	**Metric**
M_u	flexural moment demand	k-ft	kN-m
n	number, quantity	unitless	
P_n	nominal compression capacity	k	kN, MN
P_u	factored compression demand	k	kN, MN
q, q_x	area unit load, pressure	lb/ft², k/ft²	N/m², kN/m²
r	radius of a circle or cylinder	in, ft	mm, m
R	response modification factor for seismic force	unitless	
$R, R_\#$	reaction	lb or k	N, kN, MN
r_x, r_y, r_z	radius of gyration	in	mm
S	snow load	k, k/ft, k/ft² lb, lb/ft, lb/ft²	kN, kN/m, N, N/m, N/m²
s	spacing	in	mm
S, S_x, S_y	elastic section modulus	lb/in³	MN/m³
S_{DS}	seismic spectral response acceleration	unitless	
T	tension, tension force in reinforcing	lb or k	N, kN, MN
t	nominal member thickness	in	mm
T_n	nominal tension capacity	k	kN
T_u	factored tension demand	k	radians
V_n	nominal shear capacity	k	kN
V_u	factored shear demand	k	kN
w	line load, or uniform load	lb/ft	kN/m
W	wind load	k, k/ft, k/ft² lb, lb/ft, lb/ft²	kN, kN/m, kN/m² N, N/m, N/m²
W	weight	lb or k	N, kN, MN
w_D	line dead load	lb/ft	kN/m
w_L	line live load	lb/ft	kN/m
w_S	line snow load	lb/ft	kN/m

Symbol	Definition	**Imperial**	**Metric**
w_u	factored line load	lb/ft	kN/m
x	geometric axis, distance along axis	unitless	
y	geometric axis, distance along axis	unitless	
z	geometric axis, distance along axis	unitless	
ρ	reinforcing ratio		
ρ_{max}	maximum reinforcing ratio		
α	strain factor		
γ_g	grouted shear wall factor	unitless	
V_{nm}	masonry shear capacity	k	kN
V_{ns}	steel shear capacity	k	kN

Notes: 1) # indicates a general case of symbol and subscript, or subscript and symbol. It can be replaced with a letter or number, depending on how you want to use it. For example $R_\#$ may become R_L for left side reaction. Similarly, $\#_c$ may become P_c, indicating a compressive point load.

Appendix 5

Imperial to SI

Multiply		by		to get
	ft		0.305	m
	ft^2		0.093	m^2
	ft^3		0.028	m^3
	in		25.4	mm
	in^2		645.2	mm^2
	in^3		16387	mm^3
	in^4		416231	mm^4
	k		4.448	kN
	k/ft		14.59	kN/m
	k/ft^2		47.88	kN/m^2
	k/ft^3		157.1	kN/m^3
	k/in^2 (ksi)		6.895	MN/m^2 (MPa)
	k-ft		1.356	kN-m
	lb, lb$_f$		4.448	N
	lb/ft		14.59	N/m
	lb/ft^2(psf)		47.88	N/m^2 (Pa)
	lb/ft^3		0.157	kN/m^3
	lb/in^2		6894.8	N/m^2
	lb-ft		1.355	N-m
	lb$_m$		0.454	kg
	mph		1.609	kmh

SI to Imperial

Multiply		by		to get
	m		3.279	ft
	m^2		10.75	ft^2
	m^3		35.25	ft^3
	mm		0.039	in
	mm^2		0.0016	in^2
	mm^3		6.10237E-05	in^3
	mm^4		2.40251E-06	in^4
	kN		0.225	k
	kN/m		0.069	k/ft
	kN/m^2		0.021	k/ft^2
	kN/m^3		0.0064	k/ft^3
	MN/m^2 (MPa)		0.145	k/in^2 (ksi)
	kN-m		0.738	k-ft
	N		0.225	lb, lb_f
	N/m		0.069	lb/ft
	N/m^2 (Pa)		0.021	lb/ft^2 (psf)
	kN/m^3		6.37	lb/ft^3
	N/m^2		1.45E-04	lb/in^2
	N-m		0.738	lb-ft
	kg		2.205	lb_m
	kmh		0.621	mph

Appendix 5

Glossary

A615	Standard reinforcing steel grade
A706	Weldable reinforcing steel
Adhesive anchor	Post-installed anchor that transfers load through chemical bond between the adhesive, anchor, and masonry
Aggregate	Sand, gravel, crushed stone
Anchor	Steel element used to transmit load to masonry
Anchor group	Two or more similar anchors whose failure planes overlap
Area load	Load applied over an area
ASD	Allowable stress design; factors of safety are applied to the material
Axial	Action along length (long axis) of member
Axis	Straight line that a body rotates around, or about which a body is symmetrical
Balanced condition	Point when masonry crushes at same time steel yields
Beam	Horizontal member resisting forces through bending

Bed joint — Horizontal layer of mortar upon which a masonry unit is placed

Bond beam — Horizontal beam, within a wall, that is grouted and reinforced

Boundary element — Portion of wall or diaphragm edge where high tensile and compressive forces are resisted by additional reinforcement

Brace — Member resisting axial loads (typically diagonal), supports other members

Braced frame — Structural frame whose lateral resistance comes from diagonal braces

Breakout strength — Strength when a portion of masonry surrounding an anchor breaks off

Buckling — Excess deformation or collapse at loads below the material strength

Capacity — Ability to carry load, related to strength of a member

Cement — Materials that have cementing properties, such as Portland cement, fly ash, or silica fume

Chloride — Salts that accelerate reinforcing steel deterioration

Code — Compilation of rules governing the design of buildings and other structures

Collar joint — Vertical space between two wythes of masonry

Collector — See drag strut

Column — Vertical member that primarily carries axial compression load, supports floors and roofs

Component	Single structural member or element
Compression	Act of pushing together, shortening
Compression controlled point	Strain limit where masonry crushes at the same time steel reinforcement yields
Connection	Region that joins two or more members (elements)
Construction documents	Written and graphical documents prepared to describe the location, design, materials, and physical characteristics required to construct the project
Couple or force couple	Parallel and equal, but opposite forces, separated by a distance
Cover	Distance between outer edge of reinforcement and masonry edge, relates to corrosion protection of reinforcement
Creep	Slow, permanent material deformation under sustained load
Cross-sectional area	Area of member when cut perpendicular to its longitudinal axis
Crosswise reinforcement	Reinforcing running perpendicular to the long axis of the member, in the member cross-section
Dead load	Weight of permanent materials
Deflection	Movement of a member under load or settlement of a support
Demand	Internal force due to applied loads
Development length	Length reinforcement must be embedded to develop its yield (or design) strength

Diaphragm	Floor or roof slab transmitting forces in its plane to vertical lateral elements
Distributed load	Line load applied along the length of a member
Drag strut	Element that collects diaphragm shear and delivers it to a vertical lateral element
Drift	Lateral displacement between adjacent floor levels in a structure
Durability	Ability to resist deterioration
Effective height	Clear height of a member between points of lateral support
Elastic	Ability to return to original shape after being loaded
Element	Single structural member or part
Embedment	Reinforcement bar or studs welded to a plate or assembly, grouted into masonry
Empirical design	Design method based on experience, rather than theory
Extreme tension reinforcement	Reinforcement that is furthest from the extreme compression fiber
Fixed	Boundary condition that does not permit translation or rotation
Flexure, flexural	Another word for bending behavior
Force	Effect exerted on a body
Foundation	Masonry or concrete system bearing on soil, supporting structure above
Frame	System of beams, columns, and braces, designed to resist vertical and lateral loads
Free body diagram	Elementary sketch showing forces acting on a body

Girder	Beam that supports other beams
Gravity load	Weight of an object or structure, directed to the center of the earth
Grout	Mix of cement, aggregate, and water used to fill void spaces in block masonry
Head joint	Vertical layer of mortar between masonry units within a wythe
Headed bolt	Cast-in steel anchor that gains tensile strength through mechanical interlock of the bolt head or nut
Headed stud anchor	Steel anchor, with a head larger than the shaft, welded to steel plate and embedded in grout
Hoop	Closed tie with seismic hooks at each end
Inelastic	Behavior that goes past yield, resulting in permanent deformation
Lap splice	Overlapping reinforcing steel to create a splice for two bars
Layup	Pattern in which masonry is laid
Lateral load	Load applied in the horizontal direction, perpendicular to the pull of gravity
Lengthwise reinforcement	Reinforcement running along the long axis of the member
Lintel	Beam over a wall opening
Live load	Load from occupants or moveable building contents
Load	Force applied to a structure
Load combination	Expression combining loads that act together

Load factors	Factors applied to loads to account for load uncertainty
Load path	Route a load takes through a structure to reach the ground
Masonry	Construction composed of clay or cement-based bricks
Masonry strength	Specified masonry compressive strength, used in all strength calculations and correlations
Mechanical anchor	Post-installed anchor that transfers load through mechanical interlock between anchor and masonry
Modulus of elasticity	Material stiffness parameter, measure of a material's tendency to deform when stressed
Moment	Twisting force, product of force, and the distance to a point of rotation
Moment arm	Distance a force acts from a support point
Moment frame	Structural frame whose lateral resistance comes from rigid beam–column joints
Moment of inertia	Geometric bending stiffness parameter, property relating area, and its distance from the neutral axis
Neutral axis	Axis at which there is no lengthwise stress or strain, point of maximum shear stress or strain, neutral axis does not change length under load
Nominal strength	Element strength, typically at ultimate level, prior to safety factor application

P-delta	Amplification of moments due to eccentric axial loads and member or system deflection
Pier	Vertical portion of a wall, between two openings
Pin	Boundary condition that allows rotation but not translation
Point load	Concentrated load applied at a discrete location
Portland cement	Cementitious material made primarily from limestone
Post-installed anchor	Anchor installed in masonry, adhesive, or mechanical
Post tensioning	Prestressing force applied after masonry is cured
Pressure	Force per unit area
Prestressing	Compressive force applied to masonry section through steel tendons or bars to counteract tensile stresses due to gravity loads
Pretensioning	Prestressing force applied to masonry
Prism	Assembly of masonry units, mortar, and sometimes grout
Projected area	Free surface area representing an assumed failure plane starting from the masonry anchor
Pryout	Strength when masonry breaks out opposite the direction of shear in short, stiff anchors
Radius of gyration	Relationship between area and moment of inertia used to predict buckling strength

Reaction	Force resisting applied loads at end of member or bottom of structure
Reinforced masonry	Masonry with a minimum amount of reinforcing steel
Reinforcement, reinforcing	Steel embedded in masonry to carry tensile, compressive, and shear loads, and reduce thermal and shrinkage cracking (commonly A615 and A706)
Resultant	Vector equivalent of multiple forces
Rigid	Support or element having negligible internal deformation
Running bond	Placement of masonry units so head joints are at least staggered by one quarter of the unit length
Rupture	Complete separation of material
Seismic design category	Classification based on occupancy and earthquake severity
Seismic hook	Hook on a stirrup, hoop, or crosstie with 135° bend at the end
Seismic load	Force accounting for the dynamic response of a structure or element due to an earthquake
Seismic-force resisting system	Portion of structure designed to resist earthquake effects
Shear	Equal, but opposite forces or stresses, acting to separate or cleave a material, like scissors
Shear wall	Wall that resists lateral loads primarily through shear
Slab	Flat, relatively thin structural member spanning between beams and columns

Slender	Member that is prone to buckling
Slump	Measure of grout flowability, zero slump does not flow
Snow load	Load from fallen or drifted snow
Spacing	Center-to-center distance between adjacent items
Span length	Clear distance between supports
Stability	Structure's resistance to excessive deformation or collapse at loads below the material strength, opposite of buckling
Stack bond	Placement of masonry units so head joints are vertically aligned
Stiffness	Resistance to deformation when loaded
Stirrup	Reinforcement used to resist shear
Strain	Change in length divided by initial length, percent change in length if multiplied by 100
Strength	Material or element resistance to load or stress
Strength design	Load and resistance factor design; safety factors applied to the loads and materials
Strength reduction factor	Factor taking into account material strength variability
Stress	Force per unit area
Stress block	Rectangular simplification of the compressive stress in a masonry section
Structural analysis	Determination of forces, moments, shears, torsion, reactions, and deformations due to applied loads

Structural integrity	Ability of structure to redistribute forces to maintain overall stability after localized damage occurs
Structural system	Series of structural elements (beams, columns, slabs, walls, footings) working together to resist loads
Sulphate	Salts that attack the cement paste, leading to masonry deterioration
Support	Either the earth or another element that resists movement of the loaded structure or element
Tension	Act of pulling apart, lengthening
Tie	Reinforcement loop enclosing lengthwise reinforcement (typically columns)
Tributary area	Area supported by a structural member
Tributary width	Width supported by a structural member, usually a beam, joist or girt
Ultimate strength	Breaking stress, also the method in use to design masonry structures
Unbraced length	Length between brace point where a member can buckle
Veneer	Masonry wythe providing exterior finish, but has limited lateral capacity
Wind load	Force due to wind
Wythe	Continuous, vertical portion of a wall, one masonry unit thick
Yield	Point at which a material has permanent deformation due to

	applied loads, start of inelastic region of stress–strain curve
Yield strength	Specified minimum yield strength of reinforcement steel

Bibliography

ASTM, *Standard Specification for Mortar for Unit Masonry*, ASTM C270-14a (West Conshohocken: ASTM International, 2014).

"Code of Hammurabi, The" Constitution Society, accessed 11/2017. www.constitution.org/ime/hammurabi.htm.

Drysdale, R., Hamid, A. *Masonry Structures: Behavior and Design*, 3rd ed. (Longmont, CO: The Masonry Society, 2008).

Fletcher, B. *London Building Acts* (London: BT Batsford, 1914).

"History of Building Regulations," Building History, accessed 11/2017. www.buildinghistory.org/regulations.shtml.

ICBO, *Uniform Building Code* (International Conference of Building Officials, 1928, Long Beach California.

NCMA, *Section Properties of Concrete Masonry Walls*, TEK 14-1B (Herndon, VA: National Concrete Masonry Association, 2007).

OSHA, *Requirements for Masonry Construction*, 1926.706 (Washington DC: Occupational Safety and Health Administration, 2017).

Oxford English Dictionary. 2nd ed. 20 vols. (Oxford: Oxford University Press, 1989).

McMullin, P.W., Price, J.S., Persellin, E. *Concrete Design*, Architects Guidebook to Structures (New York: Routledge, 2016).

"Thatching in the City of London," Thatching Info, accessed 11/2017. http://thatchinginfo.com/thatching-in-the-city-of-london/.

TMS, *Building Code Requirements for Masonry Structures*, TMS 402-13 (Longmont, CO: The Masonry Society, 2013).

TMS, *Masonry Designers Guide*, MDG-8 (Longmont, CO: The Masonry Society, 2013).

TMS, *Specification for Masonry Structures*, TMS 602-13 (Longmont, CO: The Masonry Society, 2013).

Index

adhesive anchor, 160–162
aggregate, 26
alcoves, see cliff dwellings
alternatives, 6–7
amplification, see seismic force amplification
analysis, 144
ancestral Puebloans, 2–3, 5, 7, 13
anchor capacity, 163–181, 186–189
anchor group, 163
anchor shear, 165–169, 176–181, 186–188
anchor tension, 163–166, 169–175, 189
anchor types, 156–161
anchorage, 11, 155–191
area, 33–34, 48, 82, 84–85, 164, 192–194,
art of engineering, 20, 73, 75
assessment, xxiii–xxiv
assumptions, 78–81

balanced condition, 44–47
beam solution, 192–201
bending, 76–113
  see also flexure
block, 21–24, 53, 63
bond beam, 21–22
boundary element, 142, 152–153
brace, 77–78

breakout, 162–163
breakout strength, 165–168
brick, 16, 21–24
brick masonry, 80
buckling, 77, 115–116

calculations, xxiv
capacity, 7, 19–20, 78–86, 94–97, 102–107, 116–121, 129–134, 146, 163–181, 186–189,
cast-in-place, 156–160
cement, 21, 25–26, 52
Chartres Cathedral, 16–17
chlorides, 30–31
clay, 21
cliff dwellings, 2–5
CMU, see concrete masonry unit
code, 17–19, 57–63, 89, 146–148
collector, see drag strut
column, 21–23, 48, 51, 115–135
combined forces, 116–117
compression, 114–135
compression capacity, 116–121, 129–134
compression controlled, 44–47
compression force, 82–83
compression stress, 58–59, 66–67
compression stress block, 81–82
compressive strength, 27–28
concrete masonry, 7–8, 43, 80